Managing Power Systems

Efficient Supply Planning and Optimal Pricing of Electricity

Managing Power Systems

Efficient Supply Planning and Optimal Pricing of Electricity

NAVJOT KAUR
Punjab School of Management Studies
Panjabi University, Patiala

DEEP & DEEP PUBLICATIONS PVT. LTD.
F-159, Rajouri Garden, New Delhi - 110 027

MANAGING POWER SYSTEMS
Efficient Supply Planning and Optimal Pricing of Electricity

ISBN 81-7629-875-1

Printed in India at MAYUR ENTERPRISES,
WZ Plot No. 3, Gujjar Market, Tihar Village, New Delhi - 110 018.

Published by DEEP & DEEP PUBLICATIONS PVT. LTD.,
F-159, Rajouri Garden, New Delhi - 110 027 • Phone : 25435369, 25440916
E-mail : deep98@del3.vsnl.net.in • ddpbooks@yahoo.co.in
Showroom :
2/13, Ansari Road, Daryaganj, New Delhi - 110 002 • Telefax : 23245122

Contents

Preface

It gives me immense pleasure to author 'Managing Power Systems'. There is an increasing need for efficient supply planning and optimal pricing of electricity. Rising costs and limited investment funds have led to deferrals in capacity enhancements in India and many other developing countries of the world.

Electric utilities, a highly capital intensive sector, greatly affects the economic development of a country. In a developing economy like India, therefore, it is all the more important to integrate the electric power planning process with the national investment-planning framework and to allocate the scarce resources efficiently.

The book brings out the fact that the optimal pricing of electricity, either through real time pricing or reliability differentiated service contracts, should accurately reflect customer value of service quality, thereby helping to efficiently allocate limited supply capacity. The book also establishes within sector differentiation in value perception of reliability paving way for differentiated tariff structures.

I am very thankful to Dr. Roy Billinton, Professor of Electric Engineering, University of Saskatchewan, Saskatoon, Canada for providing guidance to me in undertaking this project. I am also indebted to Prof. Gurdeep Singh, Prof. M.S. Bedi and Prof. R.K. Sehgal for motivating me and guiding me in my study.

This book would not have been possible without the support and patience of my husband Tejpal Bhatti and son Angad.

NAVJOT KAUR

1

Introduction

In modern society the pattern of social and working habits has made mankind wholly dependent on electricity. Moreover, the society as a whole expects the electric supply to be continuously available. Electricity has great significance for people across cultures and countries. Any interruption in the supply of electricity causes not only inconvenience but also certain tangible impacts.

The basic function of a power system is to supply customers, both large and small, with electrical energy as economically as possible and with an acceptable degree of reliability and quality. But this is not always possible due to random system failures-which sometimes are beyond reasonable control of power system managers. Many a times, local disturbances, adverse weather, operational failures etc. cause the system to fail. All these failures add to the unreliability of electricity supply.

This unreliability of supply has different effects on different types of consumers. These consumers can be broadly classified into four categories—industrial, commercial, agricultural and residential. The impacts and costs of power failures to each sector are typically different from those in others.

Impact evaluation requires identification and quantification in monetary terms. Many direct impacts are relatively easy to identify and quantify, while others might be intangible (e.g. risk to life). The study of impacts of power interruptions is an important concept as it can be used in a cost-benefit analysis for developing/improving the power systems. This economic analysis is fundamental to establish a balance between expenditure of obtaining a certain level of reliability and the worth of having that level.

The present research project, is aimed at studying the impact of electricity interruptions on commercial consumers/sector in Punjab.

Commercial sector is a vital part of the economy of any state/country. Its well being is important for the well being of the entire society. Punjab is one of the most progressive states in India, having been a pioneer in green revolution. The organisation which generates and supplies electricity in Punjab is the Punjab State Electricity Board.

The Board supplies electricity to all categories of consumers in the state, having distinct classification of electricity connections and tariff structures across consumer classes. The commercial sector, (designated as non-residential supply connections) is amongst the highest tariff paying classes in Punjab.

It is assumed that electricity interruptions would have certain impact on this sector, too, as in case of other sectors. Through this research an endeavour is being made to identify and quantify that impact in both tangible and intangible terms.

For the purpose of the present research :

Impact means—effect/consequence to consumer.

Interruptions mean—any outage, planned or unplanned.

Commercial sector means—Consumers/Organisations with an electricity connection categorized as NRS (non-residential supply) by PSEB.

PUNJAB STATE ELECTRICITY BOARD—A BRIEF PROFILE

Punjab State Electricity Board (PSEB), the organisation solely responsible for generation, transmission and distribution of electricity in the state of Punjab, where this study is based, is a statutory body constituted under the Electricity (Supply) Act, 1948 subject to the provisions of the Indian Electricity Act, 1910. The Board came into existence in its present form on 1.5.1967 after the re-organization of Punjab and has since then made a remarkable contribution to the green revolution in the state apart from powering the wheel of development of various other sectors of the state. The Table 1.1 shows the growth of power sector in Punjab after the PSEB came into existence. It clearly depicits the phenomenal growth of this sector over the years with total number of consumers and connected load having risen from 6.36 lacs and 665 MW on 1.5.1967 to 46.69 lacs and 11369 MW on 31.3.1997 respectively.

The importance of providing various categories of consumers with quality electricity cannot be over emphasized as the quality of power supply, as also the quantum of per capita power consumption are key indicators to the level of development in a state. Since the study tends to determine the losses suffered by consumers as a result of interruptions in supply of power and PSEB is the sole provider of this basic need, it is pertinent to give a brief profile of the organisation.

ORGANISATIONAL SET-UP

The Board

At present the full Board consists of seven members and it is the supreme decision making authority of PSEB. Out of the seven members, six including the Chairman are Whole Time Members. Principal Secretary (Finance) to Punjab Government is an ex-officio part time member of the Board. The Board meets periodically to transact business placed before it by the Secretary Board who is responsible for presenting matters for decision by the Board, keeping a record of its decisions and for ensuring proper follow-up action from concerned quarters.

Whole Time Members

The six whole time members are collectively the second level of decision making authorities in the Board. Apart from this, each one of them has been assigned a specific area of responsibility in the Board's functioning to which he/she is

TABLE 1.1
Growth of Power Sector in Punjab

Sr. No.	Description	*As on 1.5.1967	Annual Plan ending 31.3.68	IVth Plan 31.3.74	Vth Plan 31.3.79	VIth Plan 31.3.85	VIIth Plan 31.3.90	Annual Plan ending 31.3.92	VIIIth Plan 31.3.97
1.	Installed Capacity (MW)	553	614	878	1383	2325	3050	3291	3524
2.	Generation including Import (Mus)	N.A.	1136	2282	5087	7877	15601	16702	22445
3.	Consumption including Export (Mus)	N.A.	701	1785	3687	6102	12177	13136	17744
4.	Simultaneous Maximum Demand in MW (date)	N.A.	217	484	940	1522 (7-8-84)	2771 (16-9-89)	3008 (25-6-91)	3777 (12-7-96)
5.	Length of Transmission and Distribution lines below 400 KV (Ckt. KMs)	N.A.	30713	62476	103411	158069	201384	212101	237956
6.	Transmission and Distribution Losses (%)	N.A.	—	16.95	17.12	19.03	18.96	18.76	18.00
7.	Per Capita Consumption (Kwh)	N.A.	163	196	266	377	600	610	774
8.	No. of Consumers	635562	702613	1131904	1740031	2755969	3632428	3869729	4668738
9.	Connected Load (KW)	665072	783730	1581394	2682335	4507283	6752760	7675852	11368867

Note : * Punjab State Electricity Board came into existence on 1-5-1967 after reorganisation of Punjab State.

over-all incharge. The duties of Whole Time Members are assigned by the Chairman from time to time. These functions, as assigned presently to WTM's is as given in Table 1.2.

TABLE 1.2

Punjab State Electricity Board as on 31.3.2000

1. Chairman	General Administration, Vigilance and Security Wing. Gazetted establishment (Class-I officers and above except AEE's) Issue of Show Cause notices and charge sheets to S.E.'s and Sr. Executive Engineers. Matters relating to C.E./Planning, MM, DSA, TA & QI. EA and Enf. Functions and items of work not specifically entrusted to any other Member.
2. Member-Operation	Issues relating to following Organisations : (i) All Chief Engineers/Operation (ii) Chief Engineer/RE & SIL (iii) Chief Engineer/SO & C (including NREB and Technical Committee of BBMB. (iv) Chief Engineer/Metering. (v) C.E./Store and Disposal.
3. Member-Transmission	Matters relating to following organisations : (i) Chief Engineer/Transmission lines (ii) Chief Engineer/P & M (iii) Chief Engineer/Civil Design (iv) Chief Engineer/Sub-Station (v) Chief Engineer/Work Shop (vi) Chief Architect (vii) Chief Engineer/Commercial.
4. Member Generation	Matters relating to following organisations : (i) Chief Engineer/Hydel (ii) Chief Engineer/M&MHP

(*Contd.*)

TABLE 1.2 (Contd.)

	(iii) C.E./O&M, GGSSTP, Ropar.
	(iv) C.E./O&M, GNDTP, Bathinda.
	(v) Chief Engineer/O&M, GHTP, Lehra-Mohabbat.
	(vi) C.E./Constn, GHTP, Lehra Mohabbat
	(vii) C.E./Thermal Design
	(viii) C.E./Hydel Design and Constn.
5. Member-Finance & Accounts	Finance, Budget, Revenue Expenditure, Accounts, Banking Arrangements, Loan Borrowing, Investment and Reserves, Cost Reduction, Accounts and Audit Disciplinary cases of Dy. CAO Dy. CA, Dy. FA, Sr. A.O., A.O. etc.
6. Administrative Member	Matter relating to CE/HRD. All disciplinary cases of Asstt. Engineers and Asst. Executive Engineers, Non-Gazetted Establishment of Head Office. All matters of Non-Gazetted Estt. of field offices (Except A/c's) Legal Matters, Public Relations, Language and Sports.
7. Ex-Officio (Principal Secretary-Finance)	—

ELECTRICITY STATISTICS

The following are key electricity statistics of Punjab ending 31.3.2000 (Source-MIR).

Installed Capacity and Maximum Demand (MW)

Table 1.3 gives the growth in installed capacity (MW) and maximum demand (MW).

It may be seen from above that power sector growth of Punjab has been achieved with primary emphasis being laid on steam based thermal plants while hydro power generation has not increased much all this while.

Generation (MU's), Energy Sale (MU's) and T&D Losses

Table 1.4 below gives data about power generation in MU's.

TABLE 1.3
Installed Capacity and Maximum Demand (MW)

Sl. No. Description	*1967-68*	*1984-85*	*1989-90*	*1996-97*	*1999-2000*
(A) Installed Capacity (MW)					
1. Hydro					
(a) Share from BBMB	552	1254	1254	1267	1298
(b) PSEB	48	200	516	547	547
2. Steam (PSEB)	5	860	1280	1710	2130
3. Diesel (PSEB)	9	11	—	—	—
4. Share from Central Sector	—	185	533	1014	1177
5. Grand Total	614	2510	3583	4538	5152
(B) Maximum Demand (MW)	—	1522	2771	3777	5013
Date	—	7-8-84	16-9-89	12-7-96	24-6-99

TABLE 1.4
Generation (MU's)

Sl. No. Description	*1967-68*	*1984-85*	*1989-90*	*1996-97*	*1999-2000*
1. PSEB Power Houses					
(a) Hydro	221.64	966	2664.44	3026.35	3220.17
(b) Steam	1.65	2938.27	6514.14	9778.25	13831.14
2. Share from BBMB	912.34	3592.80	4097.19	4594.79	4450.31
3. Share from Central Sector	—	380.04	2324.86	5045.66	5987.69
4. Less Auxiliary Consumption	4.67	340.59	575.50	805.43	1196.34
5. Net Energy Available	1130.96	7536.52	15025.13	21639.62	26292.97
6. Energy supplied to NFL from BBMB	1414.41	907.53	719	135.29	74.55
7. Net energy available including supply to M/s NFL	2545.37	8444.05	15744.13	21774.91	26367.52

Table 1.5 below gives data regarding sale of energy (MU's).

TABLE 1.5
Energy Sale (MU's)

Sl. No. Description	1967-68	1984-85	1989-90	1996-97	1999-2000
1. General					
(a) Domestic	79.52	654.03	1356	3082.36	3779.51
(b) Commercial (NRS)	51.74	134.20	268.95	633.59	799.19
2. Industrial	373.72	2309.03	3923.65	6765.15	7697.82
3. Other includes bulk supply, street lighting, & grid supply to MC/Agr. etc.	57.59	405.46	505.38	351.50	436.61
4. Agricultural supply	138.78	2358.99	5186.43	6342.96	8039.71
5. Theft detected	—	—	104.22	93.14	70.86
6. Underassessment on account of MCO's	—	—	44.41	52.05	55.83
7. Total sale within State	701.35	5861.71	11389.04	17320.75	20879.53
8. Export to other states	—	240.62	787.70	419.19	733.46
9. Grand Total	701.35	6102.33	12176.74	17739.94	21612.99

T&D Losses

T&D losses in PSEB have been 17.8% in 1999-2000, while the same were 37.99% during 1967-68 and nearly 19% during 1984-85 as well as 1989-90.

Data regarding Thermal Power Plants of PSEB

The plant load factor of the Thermal Plants of PSEB is as shown in Table 1.6.

Other relevant statistics pertaining to the three thermal plants are as given in Table 1.7.

Manpower Deployment

Table 1.8 shows the total manpower employed in PSEB to cater to the electricity needs of consumers in Punjab as on 31.3.2000.

TABLE 1.6

Plant Load Factor

Sl. No.	*Description*	*Installed Cap (present)*	*PLF During*			
			1984-85	*1989-90*	*1996-97*	*1999-2000*
1.	GNDTP Bathinda	4 × 110 MW	61.90	69.12	67.54	68.79
2.	GGSSTP Ropar	6 × 210 MW	59	53.84	65	74.11
3.	GHTP Lehra Mohabbat	2 × 210 MW	—	—	—	81.35

TABLE 1.7

Performance Indicators of Thermal Plants during 1999-2000

Sl. Description	*Generation cost at bus bar P/unit*	*%Plant availability*	*Man power*	*Men/KW*	*Planned outage*	*Forced outage*	*Generation*
1. GNDTP	194.84	82.9	3296	7.49	10.4	6.7	2658.8
2. GGSSTP	176.62	80.9	3466	2.75	11.1	8.0	8203.4
3. GHTP	228.25	85.9	613	1.46	6.8	7.3	2968.8

FINANCIAL STANDING OF PSEB

Table 1.9 gives the revenue a/c of PSEB during 1999-2000. During 1999-2000, PSEB suffered a commercial loss of over Rs. 1500 crore after catering for depreciation, interest etc. Thus financially PSEB is in a very delicate state.

The following, briefly are the aims and objectives of the study :

1. To Find out the Perception of Commercial Consumers Regarding Reliability of Supply

The first objective of the research would be to get an estimate of number of power interruptions from the respondents. This would be done to get an idea of the frequency and duration of electricity interruptions.

TABLE 1.8

Manpower in PSEB as on 31.3.2000

Sl. No.	Category	Sanctioned	Actual
1.	Officers		
	(a) Engg./Tech.	2805	
	(b) Gen/Admn.	174	
	(c) Accountants	101	
		3080	2797
2.	Subordinates		
	(a) Line Staff	57964	
	(b) Others	14139	
		72103	65091
3.	Class IV	13907	12028
4.	Total Regular	89090	79916
5.	Work charged & daily wagers	—	12401
	Total Manpower		92317

2. To find out the Importance of Electric Equipment to Commercial Organisations

To study the impact of electricity interruptions on consumers, it is important to establish their dependence on electricity. Thus, the objective would be to find out the criticality of electric equipments/gadgets to commercial consumers. This would establish the level of their dependence on electricity and hence the cost of unreliability of supply.

3. To find out the Extent of use of back-up Systems by Commercial Organisations

In case a consumer is critically dependent on electricity and suffers a great loss due to unreliability of supply, then some alternate arrangements would definitely be made to minimize the impact. As a sequence to the first and the second objectives of the study, i.e. to establish the level of unreliability and the dependence of consumers on electricity, the next objective

TABLE 1.9
Revenue A/c

(Rs. in crores)

A.	Revenue Receipts			
	(a) Sale of power within state			3368.57
	Sale of power outside state			139.94
	(b) Misc. receipts			381.93
	Total			**3890.44**
B.	Expenditures			
	(a) Fuel Cost	(i) Coal		1764.71
		(ii) Oil		36.67
	(b) Purchase of Power			915.45
	(c) Repairs and Maintenance			142.91
	(d) Establishment and Admn.			965.75
	(e) BBMB Exp. (Net)			14.51
	(f) Finance and other charges			31.70
				3871.7
	Gross operations surplus		(+)	18.74
	Depreciation		(-)	414.38
	Interest on instt. loans		(-)	517.14
	Lease rentals		(-)	77.62
	Interest on Govt. loans		(-)	513.52
	Commercial loss		(-)	1503.92
	Prior period adjustment		(-)	57.24
	Net Profit/loss		(-)	1561.16

would be to find out the level and extent of alternate arrangements. Generating sets are the most commonly used alternative as back-up. The aim here would be to identify the extent of use of generating systems by the commercial sector, and also the capacity of the generating sets being employed. The capability of the generating sets to run various equipments being used by the consumers would also be established.

4. To Investigate if any Health/safety risk is posed by Electricity Interruptions

One of the intangible impacts of electricity interruptions can be health/safety risk to the employees and customers of commercial organisations or to the general public. The objective would be to establish whether an interruption in electricity causes any such risk.

In cases where such a risk is established, it would be further investigated if an advance warning of the impending interruption can reduce the risk.

5. To Identify the Worst time for an Interruption to Occur

Since organisations vary in their size, location, area of operations etc., the time period when an interruption has the worst impact, may also be variable. The aim would be to identify the worst time for an interruption to occur, i.e. when the organisation would suffer the most losses due to an electricity interruption.

This would include establishing the worst month of the year, the worst day of the week and the worst time during the day for commercial organisations for an interruption to occur.

6. To Identify the Monetary Impact of Electricity Interruptions

The major impact of electricity interruptions is the monetary loss suffered by organisations as a consequence of interruptions. The aim would be to identify and estimate that monetary loss.

The estimate would be for different durations of electricity interruptions ranging from two seconds to twenty-four hours.

7. To Study the Effect of Seasonal Variations on Impact of Electricity Interruptions

The aim would be to identify any variation in the losses suffered by commercial consumers due to variation in seasons. Since Punjab has extreme seasonal variations, fluctuating between severe summers and severe winters, it would be identified if these changes bring about any change in the cost profiles of commercial consumers due to interruptions.

8. To Study the Effect of Advance Warning on Losses due to Electricity Interruptions

The objective would be to find out whether providing an advance warning about an impending interruption to the commercial consumers would help consumers in reducing their losses. It would also be established that in case the consumers are able to reduce their losses, what would be the extent of that reduction in loss and also as to how much advance warning would be required for the same.

Another objective here would be to find out whether providing information regarding the duration of interruption to the consumers just after the interruption has started would help in cost reduction. Again, information would be collected about the interruption duration, which, if intimated at the commencement of the interruption, would reduce the loss.

9. To Establish Commercial Cost Damage Function (CCDF)

The most direct and usable form of customer perceived interruption costs is customer damage function. The CDF portrays the costs associated with supply interruption as a function of outage duration.

The objective would be to find out the CCDF, and also CDF's of classes within the sector.

2

Survey of Related Literature

During the period of research various libraries/institutes in the region were visited. The available related literature in these libraries/institutes was studied which proved to be very useful in getting an insight into the main objectives of the study and in finalising the methodology. Some relevant abstracts from these studies are briefly discussed below.

In their research paper published in IEEE Transaction on Power Apparatus and Systems in Oct. 1983, G. Wacker, E. Wojczynski and R. Billinton[1] have discussed the results of a residential survey. It was carried out to study the costs incurred by residential consumers due to electricity interruptions. The study used a mail survey to determine the residential consumers' assessment of the cost of electric service interruptions. The survey obtained users' cost evaluation using three approaches. Two of these approaches asked respondents to indicate rate changes that would be commensurate with specified changes in reliability, while the third approach was an indirect worth evaluation based on the preparatory actions a consumer predicted he would take, given a specific level of unreliability. Users' evaluations are reported for a number of frequency and duration failure scenarios.

The major contribution of this work is the compilation of residential cost of interruption duration as a function of both user and interruption characteristics. It has also contributed to a great extent in developing interruption costing methodology.

The same authors,[2] in Februry 1984 issue of IEEE Transactions, present 'Interruption Cost Methodology and Results—A Canadian Commercial and Small Industry Survey'. This paper presents the results of an investigation of the direct short term impacts and costs incurred by commercial and small industrial customers resulting from local random supply interruptions. The study used a mail survey to determine the customer assessment of electric service interruptions. The variation in interruption cost with the variation in characteristics of interruptions such as duration of the interruption, time of the day, day of the week and month of the year, advance warning and reliability history and with the characteristics of the users was also investigated.

The paper has presented a summary of research conducted to determine estimates of the perceived costs or losses to commercial and small industrial consumers.

In July, 1990 issue of IEEE Transactions—Power Engineering Review, Daniel J. Ward,[3] published his paper 'Power Quality—Two different Perspectives'. This paper focuses on the problem of short duration transients or momentary disturbances and their effect on sensitive customer equipment like computers. One perspective, from the utility side of the meter, covers any power system disturbances which originate on that side of the meter and which adversely affect customers' equipment. The other perspective is from the customer side of the meter. In that regard, the causes of the disturbances are usually restricted to the customers' premises.

The author points out that many equipments like mini computers, electronic cash registers and data terminals etc. are affected because of their sensitive circuitry and cause memory loss, system malfunction and even component failure.

IEEE committee report[4] in Nov. 1990 on 'Bulk System Reliability—Predictive Indices' deals with identifying techniques and approaches for monitoring, measuring, predicting and applying reliability indices in the planning and operation of power systems.

This report identifies predictive indices that can be calculated, the degree to which they are calculated and used in practice, the commonality or lack of it between utilities, the perceived problems with their calculation or practical use, the limitations due to lack of data, computer resources and management understanding, and other related aspects.

'Design of a distribution database with customer daily load patterns', a research paper by M.Y. Cho, C.S. Chen and W.M. Lin[5] was published in Electric Power Systems Research, 1992.

In this paper, a generic database with hierarchical structure for the distribution system has been developed. The schema of the database was determined according to the data requirements to support various functions of the distribution system. This database provides distribution engineers with system information required for computer simulation. It is concluded that distribution automation can be realized through the application of a distribution database to improve the performance of distribution planning and operation. To improve the accuracy of computer simultation of the distribution system, the load patterns of various customers are derived and included in the database as one of the attributes of the customer record in the generic database. It is found that the proposed load model can represent the load behaviour quite precisely.

Outage costs, which are being estimated by the present researcher, have also been given weightage by J. Qiu and A.A. Girgis[6] in their paper 'optimization of power system reliability level by stochastic programming', published in 1993. This paper employs a stochastic programming approach for identifying a system planning scheme which leads to the optimal composite system reliability level. The objective of such approaches is to identify a system expansion scheme which leads to an optimal system reliability level corresponding to the minimal total cost, including operating and maintenance cost, expansion cost and outage cost. Benders' decomposition method is employed to decompose the optimization problem into a restricted master problem for system expansion and a sub-problem for system reliability evaluation. The sub-problem is solved by a Monte Carlo simulation considering system uncertainities. A five-bus sample system and the IEEE reliability test system have been used to test the approach and the results show that the expansion scheme obtained results in the minimal total cost.

Two comments are made by the authors—(i) As a part of the total cost, the load outage cost crucially affects the expansion scheme at the optimal reliability level. In contrast to the expansion cost which can be estimated very accurately, the load outage cost is difficult to assess accurately. Hence, the correctness of the optimal total cost will be, to a great extent, dependent on the accuracy of the load outage cost. (ii) consistency of the sub-problem solutions play an important role in the convergence of the optimization process.

Electric Power System Research (1994) carried a research paper on 'Monte Carlo simulation applied to distribution feeder reliability evaluation' authored by L. Goel and R. Billinton.[7] The authors point out that the utilization of Monte Carlo simulation techniques in reliability evaluation has increased considerably with the advent of high-speed digital computers. This paper describes a Monte Carlo approach to radial distribution feeder reliability assessment.

Goel and Billinton are of the viewpoint that reliability evaluation of individual customer service continuity levels is an important procedure which can significantly affect the design and operating characteristics of power systems in general, and distribution circuits in particular. Analysis of customer failure statistics indicates that distribution systems make a significant individual contribution to overall customer supply unavailability. Studies show that distribution systems contribute as much as 90% towards customer outages and supply unavailability. The paper is restricted to reliability evaluation of radial distribution systems.

Garry Wacker and Garry Tollefson's[8] paper 'Electric Power System customer interruption cost assessment' was published in 'Reliability Engineering and System Safety' in 1994.

In their paper, Wacker and Tollefson say that methodologies available to assess the reliability of an electric power system and the associated costs of achieving that level of reliability are well established. In contrast, the methodologies available to assess the worth to the consumer of a certain level of reliability are still evolving. The value associated with the unreliability of the electric supply system is often used as an indirect indicator of the reliability. Using this approach, it is essential to achieve some understanding of the ways that interruptions affect the customer.

The writers point out that interruption costs can be evaluated using three approaches : indirect analytical evaluations, case studies of actual blackouts and customer surveys. Results from both analytical methods and the case studies have indicated that cost assessments should obtain information that is customer specific. The best source of this information is the customers themselves. Variations of the customer survey approach appear to be growing as the method of choice for utility purposes.

Using the customer survey approach, the power systems research group has conducted a series of studies to obtain users' evaluation of worth of electric service. The authors go on to explain the customer categorization, questionnaire content, normalizing procedure of the customer survey carried out in Canada in 1991. Further cost results of residential, agricultural, commercial and industrial customers are presented with their damage functions. The four sector damage functions are described, and weighted equally to create a composite customer damage function.

L. Goel, R. Billinton and R. Gupta[9] of School of Electrical and Electronic Engineering, Nanyang Technological University, Singapore and Power System Research Group, Department of Electrical Engineering University of Saskatchewan, Saskatoon, Canada in their paper, 'Effects of adverse weather conditions and higher order outages on customer interruption costs in electric sub-transmission systems' have endeavoured to incorporate economics into the reliability cost-benefit equation. It is imperative that the cost of attaining a level of reliability be compared with the worth or benefit derived by society from that level of reliability. This paper pertains to the assessment of the value of reliability in an electric sub-transmission system. Customer outage data collected from customer survey methods are used in conjuction with system reliability indices to predict customer interruption costs that are representative of reliability worth.

The authors recognize that the basic approach to quantify the worth of electric service reliability is to estimate the monetary losses incurred by utility customers due to power supply interruptions. This has been shown to be effectively accomplished by the customer survey method. The fundamental premise on which customer surveys are based is that customers are in the best position to understand and assess the monetary implications of supply failures on their activities.

Goel, Billinton and Gupta in their paper describe that the estimation of distribution system reliability worth in terms of Supply Point Expected Interruption Cost (SPEIC) involves three basic models, the cost model, the load model and the system model. They have further described the procedure to be followed for estimating the supply point expected energy not supplied, its cost and supply point interrupted energy assessment rate at each supply point in a series of steps.

This paper presents a brief description of the failure models encountered in the analysis of electric sub-transmission systems. The concepts involved in the reliability evaluation of sub-transmission systems are illustrated by application to a relatively small test system. The basic objective of the paper is to identify and illustrate the impacts of adverse-weather and higher order failures on the worth of electric service reliability. The worth of reliability is obtained in terms of the expected interruption cost. The paper presents a failure modes and effects analysis procedure which can be used to predict the expected interruption cost for the different user sectors in a sub-transmission system. A digital computer programme, SUBTREL, was used to conduct the studies reported in the paper. The results show that adverse weather failures can contribute significantly to system reliability worth indices, while higher order outages under normal weather conditions make only a marginal contribution.

The concepts described in this paper can prove useful to system planners and managers by delineating the contributing factors to supply point reliability and customer interruption costs. Determination of the contributing factors is an important part of system design and operation.

The paper 'Power System Reliability Assessment—A Conceptual and Historical Review', by Ron Allan[10] provides a review of the conceptual aspects and historical developments relating to power system reliability, of the factors affecting it, of the underlying concepts concerning its assessment, and of the various criteria used in the different functional zones. It considers all areas of a power system including generation, transmission and distribution.

Allan points out that historically, power system reliability has been assessed using deterministic criteria, techniques and indices, and it is only recently that techniques, data and computational resources have reached the stage where probability methods can be applied to practical planning and operational decision-making.

The author discusses the functional zones of a power system (generation, transmission and distribution) as the three hierarchical levels (HL-I, HL-II and HL-III) and the probabilistic criteria and indices that can be used in planning studies.

In the same issue of Reliability Engineering and System Safety, R. Billinton and R.N. Allan[11] have discussed reliability evaluation models in the guest editorial. They point out that the basic function of a power system is to supply customers with electrical energy as economically as possible with an acceptable level of reliability including continuity and quality. This concept incorporates two opposing constraints, reliability and economics, since to increase reliability generallly means increasing investment cost. This dilemma can be solved with the interaction between the stochastic nature of system behaviour, component failures and customer demands.

Bob Simpson[12] in his paper, 'Progress towards integration of reliability concepts in a segregated electric utility system', describes the changes in the framework of the New Zealand electricity industry with particular reference to system reliability and adequacy in an evolving user pays environment.

Simpson says that Transpower is changing from engineering excellence as a business driver to one of customer service. Customers are becoming increasingly concerned over their perceived level of reliability. Outages whether they are planned or not are becoming of greater concern to them.

This paper emphasises customers' relationships with the power company to ensure a socially responsible attitude towards community impacts. Transpower is described as supplying electricity to customers on the basis of reliability to match the service paid for.

The basic function of an electric power system—to satisfy the system load and energy requirements as economically as possible and with a reasonable assurance of continuity and quality is again emphasised by Roy Billinton[13] in his 1994 paper, 'Evaluation of reliability worth in an electric power system'. This paper illustrates the application of basic power system reliability evaluation techniques to the quantification of reliability worth. The approach presented links customer interruption cost estimates with predictable indices of power system reliability.

Billinton has assigned customer cost data to the various load points in the RBTS in the form of customer damage functions. Then the individual sector CDF's have been proportionally weighted to produce a composite customer damage function. He points out that CCDF must be converted into a suitable index in order to link system reliability with the customer interruption costs. The most suitable form is known as the Interrupted Energy Assessment Rate (IEAR) expressed in \$/kwh of unsupplied energy. Description of the concepts involved in calculating an IEAR using a basic frequency and duration approach or Monte Carlo simulation is also presented. The estimation of IEAR at HL-I involves the generation of a capacity margin model which indicates the severity, frequency and duration of the expected negative margin states. This model can be used in conjunction with the CCDF for the given service area to estimate the IEAR.

At HL-II, relatively few electric power utilities apply quantitative reliability assessment techniques. The author suggests that there is a large body of published material available on HL-II reliability evaluation which describes alternate techniques and computer programmes. The author proceeds to elaborate on the estimation of distribution system reliability worth in terms of customer interruption costs. It requires a commitment to the determination of customer outage costs data in a form that can be linked with appropriate predicted reliability indices.

K. Debnath and L. Goel[14] of School of Electrical and Electronic Engineering, Nanyang Technological University, Singapore in their research paper 'Power System Planning—A Reliability Perspective', present the basic concepts of power system reliability assessment and review the reliability indices and methodologies, from a planning point of view and also from an economic cost-benefit perspective. They have tried to establish a quantitative link between the two conflicting parameters of reliability and economics of a power supply system. Such links can be established by using probabilistic criteria which consider the stochastic nature of component outages, customer demands etc.

Debnath & Goel group the three functional zones of a power system, i.e. generation, transmission and distribution into three hierarchical levels and evaluate reliability indices at each of these levels.

In the planning of a generation sub-system, the aim is to have generating capacity in excess of or at least equal to the estimated peak load. Planning is done to ensure that the probability of inadequate generation at any time is reduced to a minimum acceptable level. This minimum acceptable value is referred to as a reliability index.

The authors have prepared reliability models for generating units—the simplest being of only two states up (i.e. operating at 100% capacity) and down (i.e. forced out of service). In order to determine the value of the FOR (forced outage rates) of a generating unit, its recorded operating history for a number of years is required.

This paper also presents some dynamics associated with the determination and use of a reliability worth factor termed as the Interrupted Energy Assessment Rate (IEAR). The cost benefit approach for developing economically optimal generation standards and allocating new expenditures in transmission and distribution reliability planning is being used by several North American utilities. The essence of the cost-benefit approach lies in recognizing the fact that the total cost of electric service consists of two components, the cost of service received and the cost of service interruptions. The objective is to minimize this total cost such that the marginal cost of service reliability equals the marginal customer interruption cost.

The authors conclude that the system data, adequacy studies and cost-benefit aspects reported in this paper can be used by system planners for assessing system reliability and the economics of alternative system configurations.

Reliability Test System Task Force,[15] prepared a report on the application of probability methods, in 1996, titled The IEEE Reliability Test System-1996. This report describes an enhanced test system (RTS-96) for use in bulk power system reliability evaluation studies.

The first version of the IEEE Reliability Test System (RTS-79) was developed in 1979 by the application of probability methods sub-committee of the power system engineering committee. It was developed to satisfy the need for a standardized data base to test and compare results from different power system reliability evaluation methodologies.

Since the publication of RTS-79, several authors have reported the results of their research in the IEEE journals and many international journals using this system. Changes in the electric utility industry along with certain perceived enhancements to RTS-79 motivated the task force to suggest a multi-area RTS incorporating additional data. RTS-96, according to the committee is a hybrid and atypical system.

This paper has presented data which is required by reliability models of power systems. Future models may require other parameters.

E.H. Barakat and S.A. Al-Rashed,[16] in 1996 presented a paper on 'Analysis of the impact of dynamic events on future power requirements' in EPSR. In this paper, a model has been developed for analysing and forecasting the summer peak loads of a power system operating under conditions of high growth.

In Saudi Arabia, where the study has been carried out, high peak loads are observed during the summer period. The peak-load data during June over the period 1980-1990 was used as ex-post simulation period data in order to describe the relationship between peak load and the various explanatory factors applicable to this period. Using this relationship the peak load forecast was then produced for the ex-post forecast period 1991-1993, based on known values of the explanatory variables associated with this period.

The cyclic lunar events which affect the power demand in the country have been successfully simulated by the model. Both ex-post and ex-ante forecasts produced by the developed model compare favourably with actual loads and also with forecasts produced by an established time-series method.

The paper on 'Interruption costs, customer satisfaction and expectations for service reliability' by Michael J. Sullivan and B. Noland Suddeth[17] (U.S.A.) summarizes results of a comprehensive study of the economic value of electric service carried out by Duke Power Company, Charlotte, U.S.A. in cooperation with Electric Power Research Institute. In the study, customer interruption costs were estimated for generation, transmission and distribution outages of differing lengths occuring under varying circumstances. Interruption costs for momentary outages and voltage disturbances are also reported.

The basic methodology used in this study involves directly asking random samples of customers in different market segments (i.e. residential, commercial and industrial) to estimate their economic losses as a result of power reliability and quality problems.

Sullivan and Suddeth find that interruption costs experienced by commercial and industrial customers are much higher than residential customers. Further they conclude that interruption costs vary from customer to customer depending on a number of factors, like the customers' business type and size. Using these production factors, they developed multiple regression models for predicting customer interruption costs.

Regarding customer expectations for service reliability, the authors point out that customers differ dramatically in their expectations. Commercial and industrial customers expect nearly perfect reliability. Residential customers have significantly lower expectations for service reliability.

This study shows that customer interruption costs vary systematically and predictably as a function of customer type and size and within commercial and industrial customers by the process, equipment and products being made and sold. It also documents the effect of advance warning on interruption costs arising from generation outages and suggests that electric emergency planning may be a highly cost effective alternative to investment in new generation.

X-Liang and L. Goel[18] of Nanyang Technological University, Singapore, in 1996, published a research paper, 'Distribution System reliability evaluation using the Monte Carlo simulation method.'

The authors point out that reliability evaluation of individual customer service continuity levels is an important aspect which can significantly affect the design and operating characteristics of power systems in general, and distribution circuits in particular. Analysis of customer failure statistics indicates that distribution systems make a significant individual contribution to overall customer supply unavailability. Statistics such as these reinforce the need to conduct quantitative adequacy assessments not only for average indices but also for their distributions in the distribution functional zone.

This paper describes a Monte Carlo approach to a comprehensive distribution system reliability evaluation. The paper presents the results of load point reliability indices and system performance indices for six cases using the Monte Carlo method for a distribution test system.

In particular, the probability distributions of system performance indices such as SAIFI, SAIDI and CAIDI for the six cases are presented. The probabilities of these indices occurring during different intervals are analyzed and the effects of different operation philosophies on the system performance indices are illustrated from the probability distribution point of view. The results obtained by the simulation method are compared with those obtained using an analytical approach. This paper illustrates that Monte Carlo simulation is a useful and powerful tool for comprehensive distribution system reliability assessment. The study results show that the simulation method gives acceptable reliability indices and can also be used to provide information on the probability distributions associated with the predicted indices.

The paper on 'Reliability Planning for Composite Electric Power Systems' by Ching-Tzong Su and Guor-Rurng Lii[19] of National Chung-Cheng University of Taiwan was published in 1998 issue of EPSR. The paper proposes a reliability design approach using network flow technique, genetic algorithms and Monte Carlo simulation for composite electric power systems.

The authors say that in electric power systems, a trade off between system installation cost and power interruption cost is usually required, which gives rise to the reliability-optimization problem for electric power systems.

Evolutionary algorithms based approach for composite (generation & transmission) power systems reliability design is presented in this paper. The evolutionary algorithms are mainly based on Genetic Algorithms employing floating point representation. GA's, termed as artificial intelligence approaches, are kinds of probabilistic heuristic and stochastic algorithms that are able to find the global minimum. GA's are more flexible in solving the reliability optimization problems.

The proposed method primarily finds out the optimal values of reliability indices for the components such that the objective function composed of interruption cost and installation cost is minimized. The reliability indices adopted include Expected Demand not Served (EDNS) and Forced Outage Rate (FOR). Application of the proposed method is demonstrated using a 23 bus test system.

L. Goel, X. Liang and Y-Ou[20] of Nanyang University, Singapore in their paper 'Monte Carlo simulation-based customer service reliability assessment', published in EPSR in July '98, also emphasize the use of Monte Carlo simulation for reliability assessment.

This paper describes a MCS methodology for calculating basic load point and system performance indices and their probability distributions for distribution systems. Basic load point indices, system performance indices for four distributions of outage times by using the MCS method are presented and compared with those obtained using an analytical approach. The relative errors between the simulation and analytic methods are shown to be very small under certain specified conditions. In order to analyze the reliability performance of each type of customer, an index to calculate the probability distributions of failure and unavailability for each type has been proposed in the paper. The proposed index has been applied to a test system designated RBTS. The reliability performance of each type of customer, as obtained using the proposed index, together with the basic load point indices and the system performance indices, can be used by Public utility commissions to determine the quality of service, conduct cost/benefit analysis and even to impact rate-making decisions.

J.T. Crozier and W.N. Wisdom's[21] paper on 'A Power Quality and Reliability Index Based on Customer Interruption

Costs' is an endeavour to improve on standard reliability indexes such as SAIDI. Such indexes ignore the variation in interruption costs with customer type. Thus expenditure decisions driven by these indices that impact reliability, will not lead to the optimum improvement in value of service to customers.

The authors find that an index equal to customer interruption costs divided by energy sales overcomes this deficiency. According to Crozier and Wisdom, the costs that customers incur because of electric supply disruptions are customer interruption costs—CICs. They propose an index equal to supplier CICs divided by its energy sales in kwh. This index is equally useful for identifying best business value solutions to power quality problems. It enables a supplier to decide which portfolio of projects earns the highest return on investment, and allows a direct comparison of its value creation record with that of its competitors.

Yang Huiyi[22] of Department of Electrical Engineering and Applied Electronic Technique, Tsinghua University, China has written a paper on 'Composite system reliability evaluation based on Monte Carlo simulation combined with outages screening'. The author finds a new algorithm for composite system reliability. The algorithm is based on Monte Carlo simulation combined with predicted contingency evaluation and screening technique. Before simulation, an acceleration table is formed according to the screening of all the single and double outages, during simulation, all the multiple outages are considered and evaluated by using the acceleration table. In order to reduce the computational quantity, a heuristics algorithm is adopted. The concept of relevant regions and regulative regions are proposed based on the concept of central relaxation and are used to judge the relations between the faulted components, which make it efficient to screen possible contingencies.

The IEEE RTS (24-bus, 38-line, 32-generator) and the North-east China Power System (331-bus, 467-line, 179-generator) were evaluated by using the developed algorithm. Results show that the algorithm is credible and efficient. When the scale of the system is enlarged, the computational quantity is increased linearly. So the algorithm developed can be used for large scale system reliability evaluation. It is found, that though the probability of a multiple contingency is quite small, yet quantity of multiple contingency is large, so the cummulative probability of multiple contingencies is large enough. It is needed to consider impact of multiple contingency on composite system reliability evaluation.

Another paper by R. Billinton and Peng Wang[23] from Electrical Engineering Department University of Saskatchewan, Canada, on 'Teaching Distribution System Reliability Evaluation using Monte Carlo Simulation' was printed in IEEE transaction on Power systems in May, '99. The authors at the very outset point out that the techniques used in Power system reliability evaluation can be divided into two basic categories of analytical and simulation methods. Analytical techniques represent the system by mathematical models and evaluate the reliability indices from these models using mathematical solutions. The exact mathematical equations can become quite involved and approximations may be required when the system is complicated. A range of approximate techniques therefore have been developed to simplify the required calculations.

Billinton and Wang describe basic analytical equations to calculate load point failure rate average outage durations and average annual outage times, the convenient approach to teaching distribution system reliability evaluation. But this approach does not usually encompass any appreciation that the customer based indices can exhibit considerable natural variability on annual basis, while still having the expected values determined by the analytical approach. In order to obtain this appreciation of the annual variability, it is necessary to include probability or frequency distribution concepts in the evaluation process. The Monte Carlo simulation approach described in the paper provides the opportunity to develop an appreciation of the variability associated with the annual indices. They are of the view that increased availability of high speed computers has made Monte Carlo simulation a viable technique for many power system reliability applications.

The paper briefly illustrates the basic concepts of time sequential Monte Carlo simulation. The development of the algorithm using this technique for distribution system reliability evaluation is also described. A simulation programme is then introduced followed by a practical distribution system reliability evaluation.

The analysis of a practical test distribution system is illustrated using both analytical and simulation techniques. The analytical approach evaluates the average reliability indices using a set of mathematical equations and therefore the procedure is relatively simple and requires lesser computer time. Whereas simulation techniques evaluate the reliability indices by a series of trials and therefore the procedure is more complicated and requires longer computer time. The simulation approach can, however, provide information on the load point and system indices that the analytical techniques cannot provide.

R. Billinton and M. Pandey,[24] in their paper, 'Customer Interruption Cost Assessment : An International Perspective', published in 1999, present some interruption cost profiles obtained from quite diverse jurisdictions of the world.

The authors say that assessment of reliability has become a well established utility practice. Customer interruption costs are considered to be a surrogate for the actual worth of service reliability and widely used in its assessment.

The most common index used to link a probabilistic approach with customer interruption costs is the Expected Energy Not Supplied (EENS). This can be predicted for generation, composite generation and transmission and distribution systems using established techniques. Customer interruption cost can be expressed in terms of cost per unit of unserved energy designated as the interrupted energy assessment rate (IEAR) or the value of lost load (VOLL). The product of EENS and IEAR (VOLL) provides an explicit monetary value associated with the system condition under study and can be used in a wide range of utility decision making process.

The authors point out that though a variety of methods have been utilized to estimate the impacts of power outages on electricity consumers, the customer survey approach seems to be the most popular and practical. The most direct and usable form of customer perceived interruption costs is the customer damage functions (CDF). The CDF portrays the costs associated with supply interruptions as a function of outage duration. The costs are usually normalized by dividing the actual customer interruption costs by either the annual energy consumed or the peak load.

The paper presents the customer interruption cost profiles obtained using surveys in many countries. The results show that although the interruption costs differ in magnitude, these are large in developed countries as compared to developing countries, but the cost profiles obtained are very similar.

IEEE Transactions on Power Systems, in February 99 carried a research paper on 'Evaluation of Power systems Reliability by an Artificial Neural Network' by Nima Amjady and Mehdi Ehasan of Iran.[25]

The paper presents a new method based on Artificial Neural Networks (ANNs) for reliability evaluation of power systems. The authors say that this system can solve difficulties of the previous reliability analysis methods, such as low accuracy, complex modelling and large computations. The proposed method has been applied for reliability evaluation of generation and transmission systems. Obtained results from this method for IEEE reliability test system have been mentioned which confirm the validity of the developed approach. The proposed ANN can be easily adopted without any changes in the generation or the transmission system, since its training times are short for both systems.

Efficiency of the proposed method is dependent on the training patterns. More and better training samples can increase accuracy of this method. Also design of better training algorithms for ANNs can improve performance of this method.

In a research paper on 'Electric Service Reliability worth Evaluation for Government, Institutions and office buildings' J. Gates, R. Billinton and G. Wacker[26] of Canada present the results of an investigation conducted by Power System Research Group to determine the cost of electric service interruptions in the government, institutions and office building sector.

The authors emphasize that customers are in the best position to understand power interruption consequences in relation to their particular needs. The paper presents the results obtained through a mail survey.

The results show that the customer costs attributable to electricity supply interruptions in the GIO sector compare most closely to those found in the industrial sector.

Notes and References

1. G.Wacker, E. Wojczynski and R. Billinton, "Interruption Cost Methodology and Results—A Canadian Residential Survey", "IEEE Transactions on Power Apparatus and Systems", Vol. PAS-102, No. 10, October, 1983 (3385-3392).
2. E. Wojczynski, R. Billinton and G. Wacker, "Interruption Cost Methodology and Results—A Canadian Commercial and Small Industry Survey", IEEE Transactions on Power Apparatus and Systems, Vol. PAS-103, No. 2, Februrary 1984. (437-444)
3. Daniel J. Ward, "Power Quality—Two Different Perspectives", *IEEE Power Engineering Review,* July 1990, (66-67).
4. IEEE Committee Report, "Bulk system Reliability—Predictive Indices", *IEEE Power Engineering Review,* November 1990 (44).
5. M.Y. Cho, C.S. Chen and W.M. Lin, "Design of a distribution database with customer daily load patterns", Electric Power Systems Research 1992 (147-154).
6. J. Qiu and A.A. Girgis, "Optimization of power system reliability level by stochastic programming", Electric Power Systems Research, 1993. (87-95).
7. L. Goel and R. Billinton, "Monte Carlo Simulation applied to distribution feeder reliability evaluation", Electric Power Systems Research, 1994, (193-202).
8. Garry Wacker and Garry Tollefson, "Electric Power System Customer Interruption Cost Assessment," Reliability Engineering and System Safety 46, 1994 (75-81).
9. L. Goel, R. Billinton and R. Gupta, "Effects of adverse weather conditions and higher order outages on customer interruption costs in electric sub-transmission systems," Electric Power Systems Research, 31, 1994.
10. Ron Allan, "Power System Reliability Assessment—A Conceptual and Historical Review," Reliability Engineering and System Safety, 46, 1994 (3-13).
11. R. Billinton and R.N. Allan, "Guest Editorial", Reliability Engineering and System Safety 46, 1994 (1-2).
12. Bob Simpson, "Progress Towards Integration of Reliability Concepts in a Segregated Electric Utility System", Reliability Engineering and System Safety 46, 1994, (109-112).
13. Roy Billinton, "Evaluation of Reliability Worth in an Electric Power System", Reliability Engineering and System Safety 46, 1994 (15-23).
14. K. Debnath and L. Goel, "Power System Planning—A Reliability Perspective", Electric Power Systems Research, 34, 1995 (179-185).
15. "The IEEE Reliability Test System—1996", A Report Prepared by the Reliability Test System Task Force of the Application of Probability Methods Subcommittee, IEEE Transactions on Power Systems, Vol. 14, No. 3, August 1996.

16. E.H. Barakat and S.A. Al-Rashed, "Analysis of the Impact of Dynamic Events on Future Power Requirements", Electric Power Systems Research 37, 1996 (137-142).
17. M.J. Sullivan, B.N. Suddeth, T. Vardell and A. Vojdani, "Interruption Costs, Customer Satisfaction and Expectations for Service Reliability", IEEE Transactions on Power Systems, Vol. II, No. 2, May 1996 (989-995).
18. X. Liang and L. Goel, "Distribution System Reliability Evaluation using the Monte Carlo Simultation Method," Electric Power Systems Research, 40, 1997 (75-83).
19. Ching-Tzong Su, and Guor-Rurng Lii, "Reliability Planning for Composite Electric Power Systems", Electric Power Systems Research, 51, 1999 (23-31).
20. L. Goel, X. Liang and Y-Ou, "Monte Carlo simulation-based Customer Service Reliability Assessment", Electric Power Systems Research, 49, 1999 (185-194).
21. J.T. Crozier and W.N. Wisdom, "A Power Quality and Reliability Index based on Customer Interruption Costs," *IEEE Power Engineering Review*, April 1999 (59-61).
22. Guo Yongji, Xi Yongjian, Xiao Kai and Yang Huiyi, "Composite System Reliability Evaluation based on Monte Carlo Simulation Combined with Outages Screening", *IEEE Transactions on Power Systems*, Vol. 14, No. 2, May 1999 (785-794).
23. Roy Billinton and Peng Wang, "Teaching Distribution System Reliability Evaluation using Monte Carlo Simultation", *IEEE Transactions on Power Systems*, Vol. 14, No. 2, May 1999 (397-403).
24. R. Billinton and M. Pandey, "Customer Interruption Cost Assessment : An International Perspective", *IEEE Power Engineering Review*, January 1999.
25. Nima Amjady and Mehadi Ehsan, "Evaluation of Power Systems Reliability by an Artificial Neural Network", IEEE Transactions on Power Systems, February, 1999.
26. J. Gates, R. Billinton and G. Wacker, "Electric Service Reliability Worth Evaluation for Governments, Institutions and Office Buildings", IEEE Transactions on Power Systems, Vol. 14, February 1999.

3

Electricity Interruptions and Commercial Costs

Impact of electricity interruptions can be evaluated using three approaches—indirect analytical evaluations, case studies of actual blackouts and customer surveys.[1]

Results from both analytical methods and the case studies have indicated that cost assessments should obtain information that is customer specific. Customer-specific costs are the losses that various customers experience due to the unavailability of the functions, products and activities that are dependent upon electricity. The best source of this information is customers themselves.[2]

Therefore for the present study customer survey method was adopted. To undertake the survey the customer base was divided into appropriate major customer categories. Questionnaire was used to get information from the consumers. The detailed methodology of the study follows.

SAMPLE DESIGN

The first step in the sampling process is defining the population. For the present study, all the electricity consumers in the state of Punjab, having a commercial connection constitute the population. Information regarding both, the number of commercial connections and the total connected load (commercial) was obtained for all the cities in the state. The four cities, Jalandhar, Ludhiana, Amritsar and Patiala had the highest commercial loads, so these cities were selected for the study. These cities not only have the highest commercial load, but also the highest number of commercial connections in the whole state and constitute 40% of the population (Figures 3.1 and 3.2). Table 3.1 shows the number of commercial connections and the connected load (Commercial) in these four cities:

TABLE 3.1
Total Commercial Connections in Punjab (Population of the Study)

City	*No. of Commercial Connections*	*Comm. Connected Load (KW)*
Patiala	39,368	77,917
Amritsar	46,928	84,150
Ludhiana	64,199	1,44,245
Jalandhar	58,712	1,16,473
Total Punjab	5,33,717	9,02,633

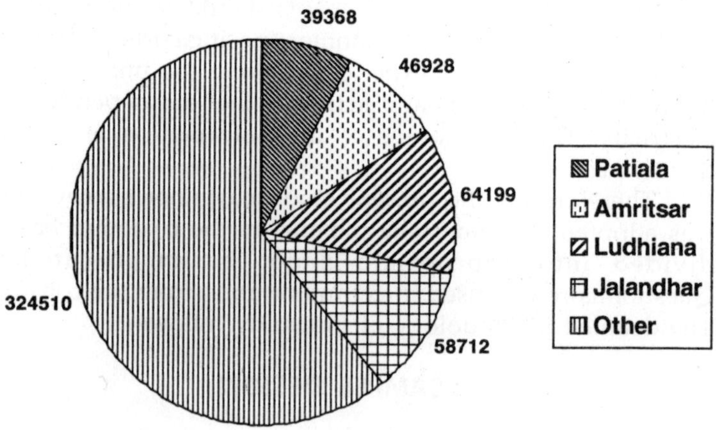

FIG. 3.1
No. of Commercial Connections

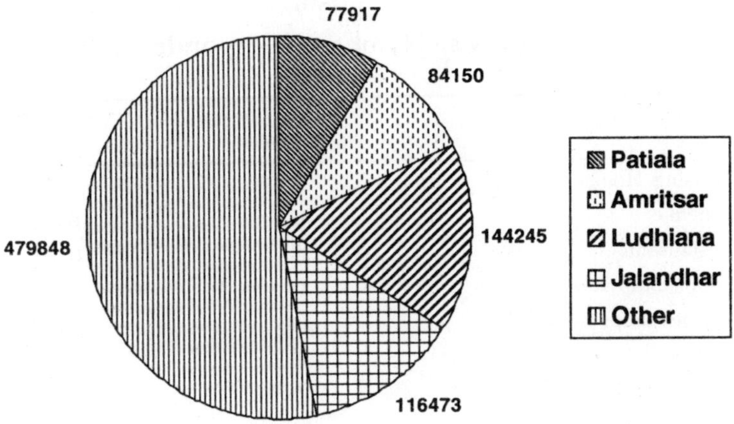

FIG. 3.2
Commercial Connected Load (KW)

In order to draw a sample, stratified random sampling technique has been used. The reason for taking a stratified sample is to have a more efficient sample than could be taken on the basis of simple random sampling. Random sampling error will be reduced, because the groups are internally homogenous, comparatively different amongst themselves. A smaller standard error may be the result of this stratified sample because the groups will be adequately represented when strata are combined. Moreover, occasionally a simple random sample yields a disproportionate number of one group or another and the representativeness of the sample could be improved by this sampling method.[3]

Size and structure is not the only characteristic but the area of operation of commercial organisations play a major role for this study. The difference in the field of operations, makes some organisations more susceptible to electricity interruptions than others. Hence to include all types of business in the sample, the population has been divided into seventeen stratas. These stratas are as per Table 3.2.

A total of 300 respondents have been taken, covering all the stratas/classes, taking 75 respondents from each city.

TABLE 3.2
Strata-wise Number of Respondents

Category	*No. of Respondents*	*%age of respondents*
Nursing Homes	19	6.33
Cinemas	20	6.67
Restaurants	30	10.00
Shops/Showrooms	27	9.00
Petrol Pumps	23	7.67
Hotels	15	5.00
Diagnostic Labs.	15	5.00
Motor Garages	14	4.67
Colour Labs.	14	4.67
Dairies	19	6.33
Marriage Palaces	15	5.00
Dry Cleaners	15	5.00
Computer Centres	17	5.67
Printing Presses	14	4.67
Chemists	16	5.33
Cable Operators	14	4.67
Beauty Parlours	13	4.33
Total	300	100

QUESTIONNAIRE DESIGN

Research tool used for the present study is a questionnaire. After a careful analysis of all the objectives of the study exercises were undertaken to construct and word appropriate questions. Many tools used by earlier researchers in the related areas were studied.

'Power Systems Research Group'[4] of Department of Electrical Engineering, University of Saskatchewan, Canada has been continuously conducting research on electricity interruptions in Canada. A questionnaire developed by the group for their study, 'Assessment of Reliability worth in Electric Power Systems in Canada' in 1993 has been adopted for the present study. This questionnaire was modified and restructured to suit Indian population/conditions.

This questionnaire was then tested for its validity and reliability.

Validity refers to the concept, that the tool actually measures what it intends to measure.

The present questionnaire's validity was measured by using the method of content validity. Face or content validity refers to the subjective agreement among professionals that a scale logically appears to be accurately reflecting what it proposes to measure. And that the content of the scale appears to be adequate.[5] When it appears evident to experts that the measure provides adequate coverage of the concept, the measure has face validity. So, the questionnaire was shown to various academicians and also other experts for checking its content validity. A number of modifications were made on the basis of the suggestions received from these experts.

Reliability refers to the concept that when the outcome of the measuring process is reproducible, the measuring instrument is reliable. Reliability applies to a measure when similar results are obtained over time and across situations. The two dimensions underlying the concept of reliability are : repeatability and consistency.[6] To measure the reliability of the present scale, the test-retest method was used. It involves administering the same scale to the same respondents at two separate points in time to test for stability. The two sets of scores obtained are then checked for their correlation. In the present case a correlation score of .90 was obtained when the scale was administered to the same respondents at two different points of time.

Hence questionnaire was found to be valid and reliable to be used for the present research project.

COLLECTION OF DATA

Survey-also known formally as a sample survey method of data gathering has been adopted. Surveys have an advantage over other methods, as these provide a quick, inexpensive, efficient and accurate means of gathering information about the population. A review of the literature reveals that investigations into the cost of power interruptions have been conducted using a variety of approaches.[7] The approach which is considered to yield relatively definitive results in the assessment of direct costs, particularly for such customers as commercial and industrial users, is the customer survey approach.[8] Hence it was decided to use survey to conduct the study. The basis of this approach is that customers are in the best position to understand power interruption consequences in relation to their particular needs.

Every respondent in the present sample was contacted personally to get the questionnaires filled. Personal contact with the respondents facilitates the collection of accurate data, as any misunderstanding on the part of the respondent is cleared by the researcher on the spot.

Data for the present project was collected through personal contact during the summer months of the year 1999. The respondents were contacted during the day, at their place of work. Subsequent visits were made if questionnaire could not be filled during one visit due to busy schedule of the respondents. In case of all the 300 respondents either owner/ partner or manager was asked to answer the questions to ensure the accuracy and reliability of the information.

DATA ANALYSIS

To analyse the data, in order to get meaningful information (results) regarding the impact of electricity interruptions on commercial consumers, mainly percentages, averages have been calculated. A particular effect of an interruption has been studied as to how much percentage of respondents have been affected.

Direct Costs

Direct costs estimated by respondents have been calculated in a similar fashion. Costs indicated by respondents for specific durations of interruptions have been averaged for all the categories.

Cost Normalization

The primary purpose of conducting this study is to establish monetary losses associated with various levels of unreliability. These costs must be in a form that is most usable to the organisation supplying electricity. A simple average or mean value of interruption costs can create a distorted understaning of the actual costs associated with an interruption. Therefore the costs have been normalized by using the connected loads of the consumers. The procedure used to normalize costs is an 'aggregated average' approach. The interruption costs associated with the respondents are totalled and the respondents' connected loads are also totalled. The ratio of the sum of the costs and the sum of the connected loads is designated as an 'aggregated average'. This procedure produces a different result than if the individual users' costs were normalized and then averaged.[9] The aggregated averages approach is used to off-set the impact of small numbers of large or small consumers, or small numbers of respondents who report large or small costs.

COMMERCIAL COST DAMAGE FUNCTION

The most direct and usable form of customer perceived interruption costs is the customer damage function.[10] It portrays the costs associated with outages as a function of outage duration. The CDF can be determined for a particular customer category or for an overall sector. Commercial Cost Damage Function has been calculated for the sector and also for classes within the sector by weighting the aggregated customer costs by the percentage of their load mixes and then summing these costs.

Notes and References

1. Wacker, G. and Billinton R., "Customer Cost of Electric Interruptions", IEEE Proc. 77(6), 1989.
2. Garry Wacker and Garry Tollefson, "Electric Power System Customer Interruption Cost Assessment", Reliability Engineering and System Safety 46, 1994.
3. William G. Zikmund, "Exploring Marketing Research", The Dryden Press, CBS Publishing Japan Ltd., 1986.
4. Power System Research Group, "Assessment of Reliability Worth in Electric Power Systems in Canada", Department of Electrical Engineering, Saskatoon, June 1993.

5. William G. Zikmund, "Exploring Marketing Research", The Dryden Press, CBS Publishing Japan Ltd., 1986.
6. William G. Zikmund, "Exploring Marketing Research", The Dryden Press, CBS Publishing Japan Ltd., 1986.
7. G. Wacker, E. Wojczynski, R. Billinton, "Cost/Benefit Considerations in Providing an Adequate Electric Energy Supply," Third Symposium on Large Engineering Systems, July 1980.
8. E. Wojczynski, R. Billinton and G. Wacker, "Interruption cost Methodology and Results—A Canadian Commercial and Small Industry Survey", IEEE Transactions on Power Apparatus and Systems, Vol. PAS-103, No. 2, Feb. 1984.
9. Garry Wacker and Garry Tollefson, "Electric Power System Customer Interruption Cost Assessment", Reliability Engineering and System Safety (46) 1994.
10. R. Billinton and M. Pandey, "Customer Interruption Cost Assessment : An International Perspective", *IEEE Power Engineering Review*, Jan. 1999.

4

Impact of Electricity Interruptions

To study the impact of electricity interruptions, the respondents' experience with electricity supply service was sought. This immediately established the context for impending questions, since as the respondents thought about the number of interruptions they experienced, they subsequently began visualising the effect those interruptions had on their business.

The next array of questions tended to evaluate the specific effects an interruption had on the activities going on/ equipment used by the consumers in their organisations. This also established the importance of certain equipment, negative effects an interruption had on the business and the role of backup systems employed.

Thereafter the worst period for an interruption was identified. This led to estimation of interruptions costs in monetary terms due to an interruption during the worst period as well as seasonal variations therein. The data was collected personally and special care was taken to overcome the various difficulties encountered by respondents in quantifying the effect of power interruptions.

NUMBER OF INTERRUPTIONS

All the 300 respondents were asked to report the number

of interruptions in electricity that occured in the last one year. They were asked to further state how many of those interruptions lasted for four hours or more. This information would indicate the magnitude of losses that the population had to suffer due to these interruptions.

Table 4.1 shows the number of failures reported by respondents in last one year. The number of interruptions which lasted for four hours or more are indicated in Table 4.2.

TABLE 4.1
Number of Failures in Last One Year

No. of Failures	*No. of Respondents*	*%age of Respondents*
0	0	0
1-100	95	31.67
101-200	84	28.00
201-300	45	15.00
301-400	20	6.67
401-500	16	5.33
501-600	12	4.00
More than 600	28	9.33
Average No. of failures	292.75	

The breakdowns/failures reported by the respondents indicate that during the year the average number of interruptions reported are quite high - 292.75 (say 293). Majority of the respondents that is 31.67% (95 respondents) reported failures between 1 to 100. 28% (84) respondents indicated the number of interruptions to be between 101 and 200. 15% (45) respondents estimated the interruptions to be more than 200 but less than 300. About 1/4th of the respondents (76) reported the interruptions to be more than 300 in number. This clearly indicates that the number of electricity failures reported by the commercial respondents is quite high which gives an impression of an unreliable service.

As is shown in Table 4.2, only 5 respondents, i.e. 1.67% reported no failure which lasted for more than four hours. A

TABLE 4.2
Number of Failures Lasting four hours or more in last one year

No. of Failures	*No. of Respondents*	*%age of Respondents*
0	5	1.67
1-10	135	45.00
11-20	73	24.33
21-30	36	12.00
31-40	9	3.00
41-50	32	10.67
51-60	3	1.00
61 or more	7	2.33
Average No. of failures	21.65	

large majority, i.e. 208 respondents (69.33%) reported such interruptions to be upto 20 in number, although only 2.33% respondents reported these interruptions to be more than 60 in number. Tables 4.1 and 4.2 clearly show that the number of electricity interruptions in Punjab are quite high and quite a few of these are of a relatively long duration. The average number of failures lasting 4 hours or more during the year are 21.65 (say 22).

IMPORTANCE OF ELECTRIC EQUIPMENT

Electricity is of paramount importance for successful running of a business as most modern day commercial organizations are critically dependent upon electric gadgets/ equipments for their day to day functioning. The impact of electricity interruptions would definitely be more if the resultant non functioning of critical equipment hampers the working of the organisation. So to study the impact of electricity failures and the losses these cause it is important to find out the type of electric equipments used by the reporting organisations and also their importance to the business.

Table 4.3 indicates the type of electric equipment used and their level of importance to the business houses.

TABLE 4.3
Importance of Electric Equipment

Electric Equipment	Do not have		Not at all Imp.		Not very Imp.		Quite Imp.		Very Imp.	
	N	%	N	%	N	%	N	%	N	%
Air Conditioning	102	34.00	10	3.33	32	10.67	18	6.00	138	46.00
Water Heating/Cooling	148	49.33	20	6.67	40	13.33	41	13.67	51	17.00
Ventilation Equipment	26	8.67	11	3.67	44	14.67	67	22.33	152	50.67
Office Equipment	175	58.33	10	3.33	61	20.33	28	9.33	25	8.33
Electric Cash Register	282	94.00	3	1.00	5	1.67	4	1.33	6	2.00
Computers	171	57.00	3	1.00	7	2.33	11	3.67	108	36.00
Telecom Equipment	119	39.67	20	6.67	75	25.00	45	15.00	41	13.67
Cooking Equipment	270	90	1	0.33	3	1.00	7	2.33	19	6.33
Motors	257	85.67	0	0	0	0	5	1.67	38	12.67
Pumps	179	59.67	1	0.33	1	0.33	18	6.00	101	33.67
Refrigeration and Freezing	163	54.33	0	0	8	2.67	8	2.67	120	40.00
Elevators	272	90.67	0	0	0	0	5	1.67	23	7.67
Indoor Lighting	0	0	2	0.67	13	4.33	49	16.33	236	78.67
Outdoor Lighting	0	0	5	1.67	26	8.67	127	42.33	142	47.33
Any other	158	52.67	0	0	0	0	1	0.33	141	47.00

Respondents were asked to rate the importance to their organisations of various types of equipment that run on electricity. They could choose from five options : 'do not have', 'not at all important', 'not very important', 'quite important' and 'very important'.

The equipment that is reported to be very important indicates that the working of the organisation is dependent on the equipment for its operations.

Air conditioning has been reported to be very important by 46% respondents (Fig. 4.1). Another 6% report air-conditioning to be quite important. Therefore for more than half the respondents, airconditioning has vital importance.

Another equipment which is reported to be very important by a majority of respondents is ventilation. 50.67% report

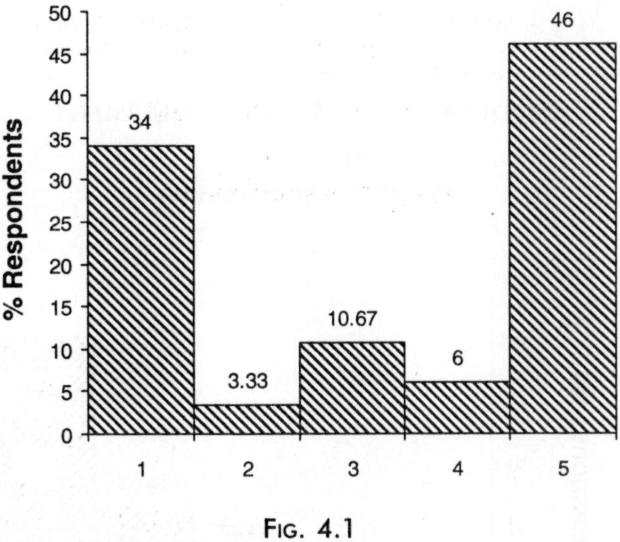

FIG. 4.1
Air-Conditioning

ventilation which includes exhaust fans etc., to be very important and another 22.33% to be quite important. (Fig. 4.3)

Computers also rate as very important for 36.00% respondents and quite important for 3.67% respondents (Fig. 4.6). Refrigeration and freezing is also reported to be very important for 40% of the commercial respondents (Fig. 4.7).

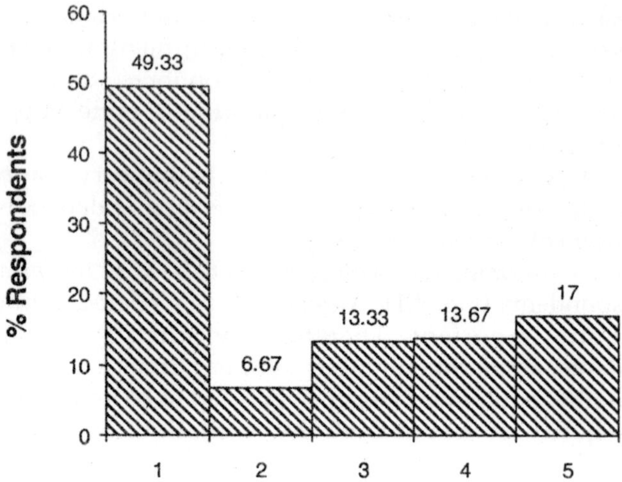

1 — Do not have
2 — Not at all important
3 — Not very important
4 — Quite important
5 — Very important

IMPORTANCE OF ELECTRIC EQUIPMENT

FIG. 4.2
Water Heating/Cooling

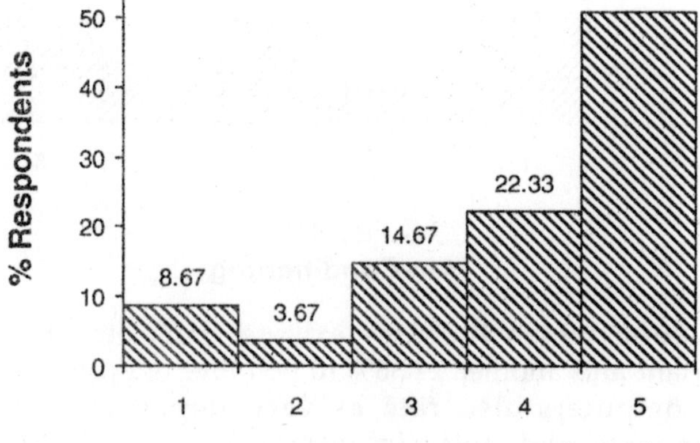

FIG. 4.3
Ventilation Equipment

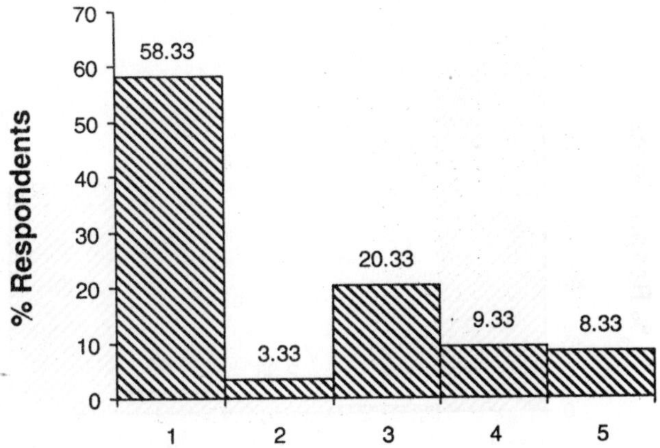

1 — Do not have
2 — Not at all important
3 — Not very important
4 — Quite important
5 — Very important

IMPORTANCE OF ELECTRIC EQUIPMENT

FIG. 4.4
Office Equipment

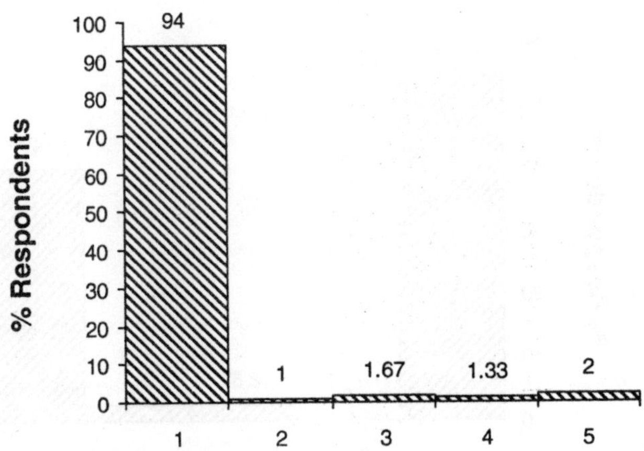

FIG. 4.5
Electric Cash Register

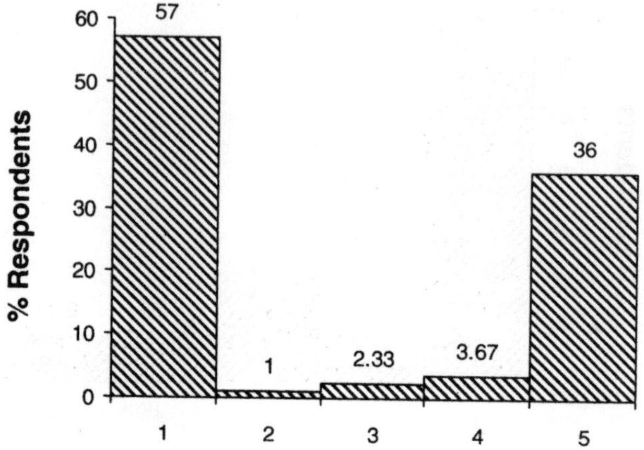

1 — Do not have
2 — Not at all important
3 — Not very important
4 — Quite important
5 — Very important

IMPORTANCE OF ELECTRIC EQUIPMENT

FIG. 4.6
Computers

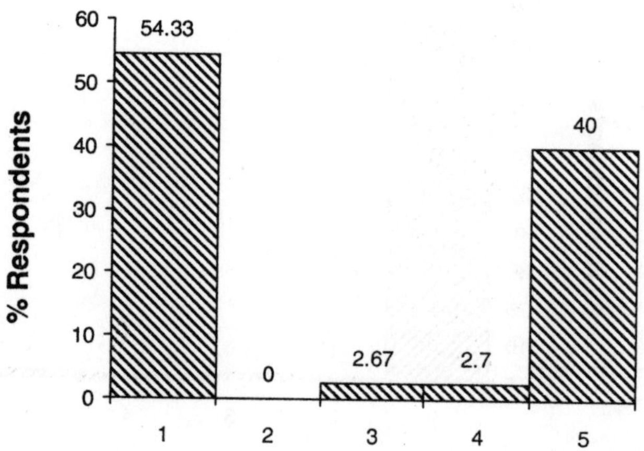

FIG. 4.7
Refrigeration and Freezing

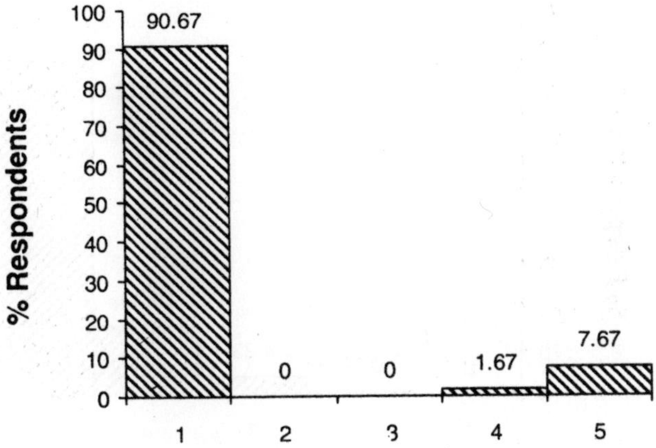

1 — Do not have
2 — Not at all important
3 — Not very important
4 — Quite important
5 — Very important

IMPORTANCE OF ELECTRIC EQUIPMENT

FIG. 4.8
Elevators

Indoor and outdoor lighting also rate as very important for majority of respondents. Indoor lighting has been reported as very important for 78.67% respondents and quite important for 16.33% respondents, making it a total of 95% of respondents (Fig. 4.9). Hence it comes out to be the most important equipment. Outdoor lighting is also very important for 47.33% respondents and quite important for 42.33% respondents, making it to be 89.66% (Fig. 4.10).

An effort was made to cover all types of electric equipment. However some specialised equipment which have not been listed, are covered under 'any other' category by the respondents. In this category 47% respondents have responded as equipment being very important (Fig. 4.15). The equipment in 'any other' category include special medical equipment, machinery used by printing presses, dry cleaners etc.

Therefore, as indicated by Table 4.3, equipment like air conditioning, ventilation, refrigeration, indoor and outdoor lighting and some specialized equipment come out to be the

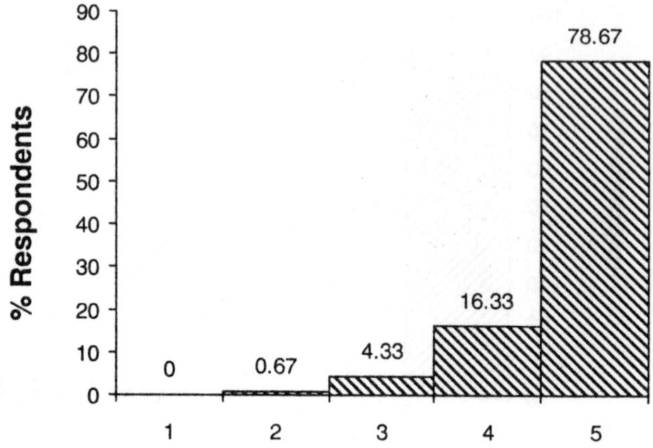

FIG. 4.9
Indoor Lighting

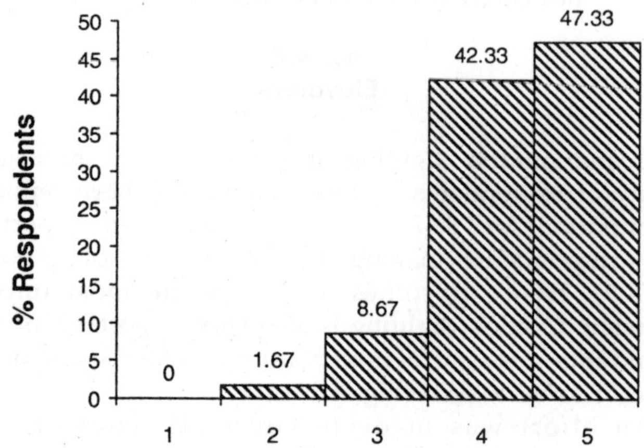

1 — Do not have
2 — Not at all important
3 — Not very important
4 — Quite important
5 — Very important

IMPORTANCE OF ELECTRIC EQUIPMENT

FIG. 4.10
Outdoor Lighting

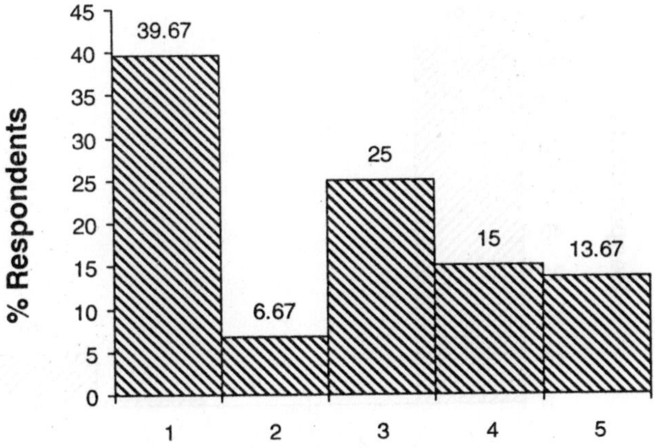

FIG. 4.11
Telecom Equipment

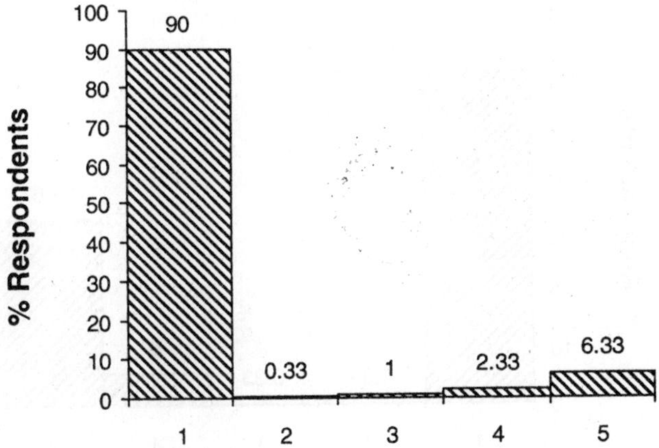

1 — Do not have
2 — Not at all important
3 — Not very important
4 — Quite important
5 — Very important

IMPORTANCE OF ELECTRIC EQUIPMENT

FIG. 4.12
Cooking Equipment

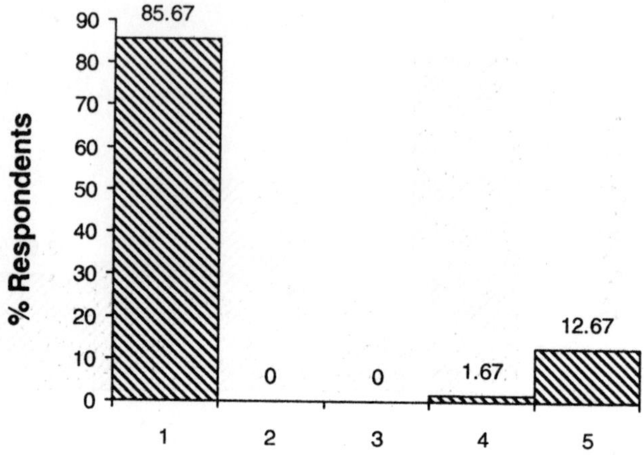

FIG. 4.13
Motors

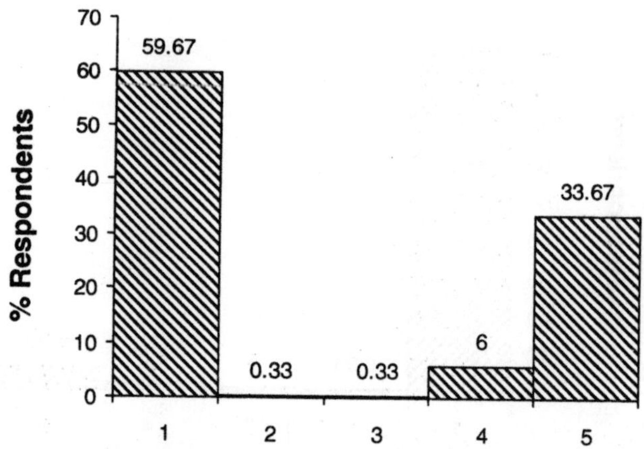

1 — Do not have
2 — Not at all important
3 — Not very important
4 — Quite important
5 — Very important

IMPORTANCE OF ELECTRIC EQUIPMENT

FIG. 4.14
Pumps

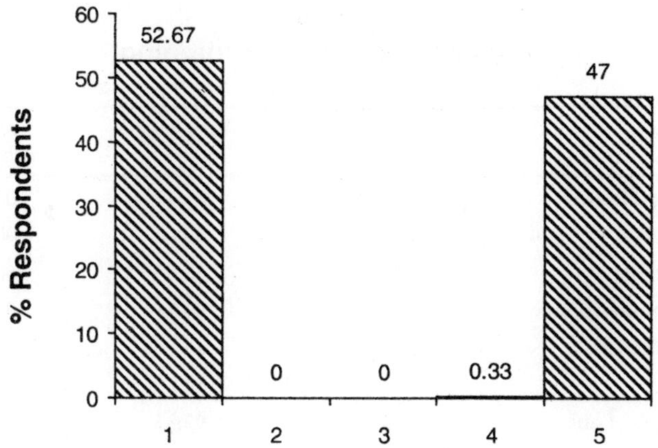

1 — Do not have
2 — Not at all important
3 — Not very important
4 — Quite important
5 — Very important

IMPORTANCE OF ELECTRIC EQUIPMENT

FIG. 4.15
Special Equipment

most important to commercial consumers. The responses indicate that the total operation of certain types of business depend wholly on these equipment and in these cases, electricity interruption results in total stoppage of the equipment which in turn has a major impact.

Table 4.4 shows that for certain types of business like hotels, nursing homes, cinemas, computer centres etc., air conditioning is of utmost importance.

Likewise, ventilation is very important for a large number of business operations. For businesses like cinemas, restaurants, shops, hotels, motor garages, printing presses, beauty parlours etc. ventilation is of high importance.

Refrigeration and freezing comes out to be very important for businesses like nursing homes, restaurants, hotels, diagnostic labs, dairies and chemists who report their business to be largely dependent on refrigeration and freezing, as indicated by Table 4.5.

TABLE 4.4
Importance of Airconditioning

Category	*Quite Imp.* %	*Very Imp.* %	*Total for which Imp.%*
Nursing Homes	10.53	89.47	100
Cinemas	5	85.00	90
Restaurants	—	60.00	60.00
Shops/Showrooms	—	59.26	59.26
Petrol Pumps	4.35	—	4.35
Hotels	—	100	100
Diagnostic Labs	—	93.33	93.33
Motor Garages	14.29	—	14.29
Colour Labs	21.43	50.00	71.43
Dairies	—	—	—
Marriage Palace	13.33	6.67	20
Dry Cleaners	—	6.67	6.67
Computer Centres	11.76	88.24	100
Printing Presses	14.29	35.71	50
Chemists	6.25	—	6.25
Cable Operators	14.29	—	14.29
Beauty Parlours	—	92.31	92.31

Indoor and outdoor lighting is also very important for most of the business organisations. Infact it is the single most important equipment for all categories of businesses, as Table 4.6 indicates.

Indoor lighting is reported to be important by 100% of respondents in all business areas except petrol pumps, dairies and cable operators where importance is indicated as 87%, 68% and 57.14% of respondents respectively.

Likewise, in case of outdoor lighting, barring cable operators, all other respondents have given great importance to outdoor lighting.

47% respondents have chosen 'any other' equipment as very important. Where ever the business operation is dependent on any special electrical equipment, this response has been

TABLE 4.5

Importance of Refrigeration and Freezing

Category	*Quite Imp.* %	*Very Imp.* %	*Total for which Imp.%*
Nursing Homes	10.53	84.21	94.74
Cinemas	10.00	20.00	30.00
Restaurants	—	100.00	100.00
Shops/Showrooms	—	14.81	14.81
Petrol Pumps	—	—	—
Hotels	13.33	86.67	100.00
Diagnostic Labs	6.67	86.67	93.34
Motor Garages	—	—	—
Colour Labs	—	—	—
Dairies	—	100.00	100.00
Marriage Palaces	—	6.67	6.67
Dry Cleaners	—	—	—
Computer Centres	—	—	—
Printing Presses	—	—	—
Chemists	—	93.75	93.75
Cable Operators	—	—	—
Beauty Parlour	7.69	38.46	47.15

chosen. These equipment include medical equipment like E.C.G. machines, X-ray equip. etc. in case of nursing homes, projectors in case of cinemas, drycleaning m/c in case of drycleaners, printing and paper cutting m/c in case of printing presses and some special equipment in case of beauty parlours. Thus, nearly half of the respondents do depend on electricity for their major area of business and unreliability of supply cripples their operations.

All the 300 (100%) of respondents have indicated great importance of one electric equipment or the other. This means that electricity assumes vital importance for all types of commercial consumers. At the same time, these respondents have indicated a significant level of unreliability in the electricity supply, shown by the number of electricity interruptions in Tables 4.1 and 4.2.

TABLE 4.6
Importance of Lighting (Indoor and Outdoor)

Category	*Indoor Lighting Total for which Imp.%*	*Outdoor Lighting Total for which Imp.%*
Nursing Homes	100	89.47
Cinemas	100	100
Restaurants	100	93.33
Shops/Showrooms	100	96.30
Petrol Pumps	87	100
Hotels	100	86.67
Diagnostic Labs	100	73.34
Motor Garages	100	100
Colour Labs	100	100
Dairies	68	94.74
Marriage Palaces	100	100
Dry Cleaners	100	100
Computer Centres	100	88.24
Printing Presses	100	85.71
Chemists	100	100
Cable Operators	57.14	14.29
Beauty Parlours	100	76.92

ROLE OF BACK UP SYSTEM

Since all the respondents have indicated unreliability of supply and also their dependence on electric equipment, they were further asked if they had any back-up system in case of electricity interruptions. There are a variety of back-up systems available, the more popular being engine generater system and battery based system (invertors etc.). Both have a significant initial cost and regular maintenance costs. The engine system has a substantial running cost in the form of the fuel consumed.

Inspite of these costs associated with back-up systems, 20% of respondents have both engine and battery systems, 3.33% of respondents have only battery systems, 64.67% of respondents have only engine generating systems and the rest 12% have no back up. This significantly high number of respondents making use of back-up systems as shown in Table 4.7 and Fig. 4.16 indicates the vital nature of the service and also its unreliability.

TABLE 4.7
Back-up System

Category	*Battery Sys.*		*Engine Generators*		*Both*		*No back-up*	
	N	%	N	%	N	%	N	%
Nursing Homes	1	5.26	14	73.68	4	21.05	—	—
Cinemas	—	—	18	90.00	2	10.00	—	—
Restaurants	1	3.33	25	83.33	2	6.67	2	6.67
Shops/Showrooms	—	—	20	74.07	4	14.81	3	11.11
Petrol Pumps	—	—	16	69.57	7	30.43	—	—
Hotels	—	—	12	80.00	3	20.00	—	—
Diag. Labs.	—	—	10	66.67	4	26.67	1	6.67
Motor Garages	—	—	10	71.43	4	28.57	—	—
Colour Labs.	1	7.14	8	57.14	2	14.29	3	21.43
Dairies	1	5.26	5	26.32	—	—	13	68.42
Marriage Palaces	—	—	12	80.00	3	20.00	—	—
Dry cleaners	—	—	10	66.67	1	6.67	4	26.67
Computer Centres	2	11.76	6	35.29	9	52.94	—	—
Printing Presses	—	—	4	28.57	5	35.71	5	35.71
Chemists	—	—	12	75.00	—	—	4	25.00
Cable Operators	2	14.29	3	21.43	9	64.29	—	—
Beauty Parlours	2	15.38	9	69.23	1	7.69	1	7.69
Total	10	3.33	194	64.67	60	20.0	36	12.0

Respondents were also asked to indicate the type of equipment that these back-up systems served. As shown in Table 4.8, 38.67% of the respondents have to make use of generating systems, which can serve all their equipment, since running of all their equipment is essential for their operations.

The major types of businesses which make use of generating systems serving all their equipment include cinemas, petrol pumps, marriage palaces and cable operators etc. Battery

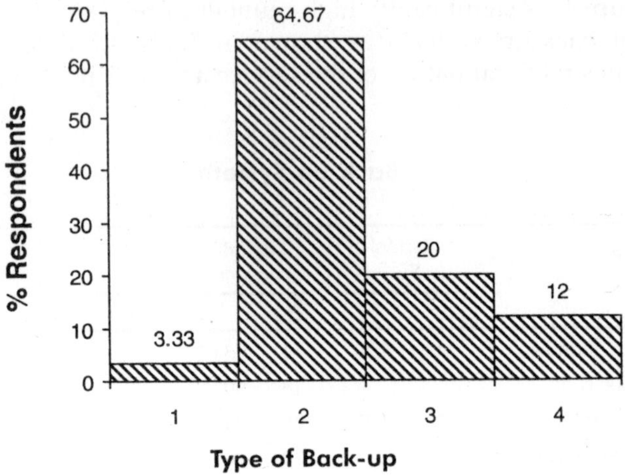

1 — Battery System
2 — Engine Generator
3 — Both-Battery System and Engine Generator
4 — No Back up

FIG. 4.16
Back up System

TABLE 4.8
Equipment Served

Type of Back-up	*Some Equip.* %	*All Equip.* %
Engine Gen.	61.33	38.67
Battery Sys.	100	—

system is made use of only to operate lights and fans or computers etc. Where ever a generating system is required to run all equipment, engine generating systems are used, though it also involves expenditure on fuel on a regular basis. The categoreis spending more on fuel for generating systems are nursing homes, cinemas, hotels and marriage palaces as these cannot operate without some of their equipment. These are also the business categories making use of generators with large capacities, the average capacity in the commercial sector being

25.23 kw. For business areas like cinemas and hotels the capacity in some cases is as high as over 100 kw. This makes these categories spend heavily on generating systems and regularly on fuel. So these are identified as being more affected by unreliability of supply.

HEALTH/SAFETY HAZARDS DUE TO ELECTRICITY INTERRUPTIONS

Apart from this ineviTable loss to commercial sector, where these organisations have to spend on generating systems and fuel, another major impact of interruptions can be in the form of health or safety hazard when there is a sudden electricity interruption. Respondents were asked to indicate if failure in electricity supply resulted in any health or safety hazard to their employees or to public and would this risk be reduced if a warning is given prior to an interruption, in case of planned interruptions, during load shedding or maintenance work.

Respondents were provided with four choices i.e. 'greatly', 'moderately', 'slightly' and 'not at all'. Here respondents had to indicate if an electric interruption would pose a safety or health risk to their customers, employees or public.

As shown by Table 4.9 and Figure 4.17 some types of business organisations foresee sudden electric interruptions to be risky. Such organization include nursing homes, where only 1 (5.26%) respondent has responded as electric interruptions not at all causing any safety risk. Rest of the respondents (94.74%) find such interruptions causing risk and a majority, i.e. 63.16% find the risk to be great. This is because of certain medical procedures going on in nursing homes and the life of patients being involved.

Similarly in case of diagnostic labs, too only 13.33% respondents do not find any safety hazard, the rest 86.67% finding such interruptions to be hazardous to health/safety because a power failure can interrupt the testing procedure or cause error in results.

Other categories of business indicating such risk are cinemas and hotels, where the risk is because of presence of large numbers of people in the premises and in case of sudden interruptions, possibility of stampede, spoilage of perirhables

TABLE 4.9
Health/Safety Hazard Caused by Electricity Interruptions

Category	*Greatly*		*Moderately*		*Slightly*		*Not at all*	
	N	*%*	*N*	*%*	*N*	*%*	*N*	*%*
Nursing Homes	12	63.16	5	26.32	1	5.26	1	5.26
Cinemas	—	—	6	30.00	10	50.00	4	20.00
Restaurants	1	3.33	1	3.33	—	—	28	93.33
Shops/Showrooms	—	—	1	3.70	—	—	26	96.30
Petrol Pumps	—	—	—	—	—	—	23	100.00
Hotels	—	—	3	20.00	8	53.33	4	26.67
Diag. Labs.	1	6.67	5	33.33	7	46.67	2	13.33
Motor Garages	—	—	—	—	—	—	14	100.00
Colour Labs.	—	—	—	—	—	—	14	100.00
Dairies	—	—	1	5.26	—	—	18	94.74
Marriage Palaces	—	—	2	13.33	6	40.00	7	46.67
Dry Cleaners	—	—	—	—	—	—	15	100.00
Computer Centres	—	—	—	—	—	—	17	100.00
Printing Presses	—	—	—	—	1	7.14	13	92.86
Chemists	4	25	11	68.75	1	6.25	—	—
Cable Operators	—	—	—	—	—	—	14	100.00
Beauty Parlours	—	—	—	—	—	—	13	100.00
Total	18	6	35	11.67	34	11.33	213	71

etc. being there. From the total sample, 29% respondents have indicated some health/safety hazard involved with interruptions without warning and 6% find this risk to be great.

ROLE OF ADVANCE WARNING TO REDUCE RISK

The respondents who indicated some risk/hazard in case of interruptions were asked to indicate as to how much warning period was required to reduce the risk. As threat to life is a major impact of electricity interruptions, it becomes the duty of the power utility, (PSEB) to help reduce this risk wherever

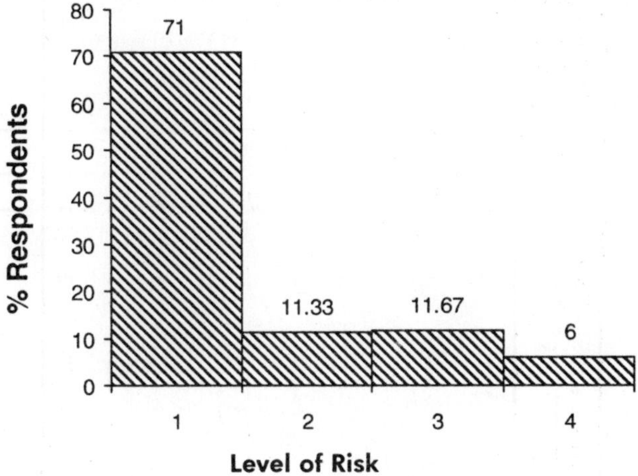

1 — Not at all
2 — Slight risk
3 — Moderate risk
4 — Great risk

FIG. 4.17
Health/Safety Hazard

possible. The shortest warning periods required to reduce the risk are indicated in Table 4.10 and Figure 4.18.

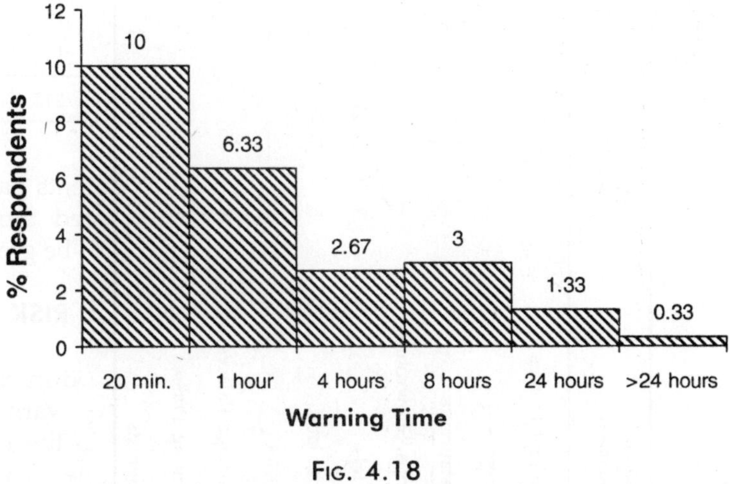

FIG. 4.18
Warning Time to Reduce Health/Safety Hazard

TABLE 4.10
Warning Time to Reduce Risk

Category Equipment	*20 minute*		*1 hour*		*4 hours*		*8 hours*		*24 hours*		*>24 hours*	
	N	%	N	%	N	%	N	%	N	%	N	%
Nursing Homes	3	16.6	5	27.7	4	22.2	5	27.7	—	—	1	5.26
Cinemas	11	68.75	3	18.75	1	6.25	1	6.25	1	6.25	—	—
Restaurants	—	—	—	—	1	50.00	—	—	—	—	—	—
Shops/Showrooms	—	—	1	100	—	—	—	—	—	—	—	—
Hotels	3	27.27	4	36.36	—	—	—	—	—	—	—	—
Diag. labs.	5	38.46	2	15.38	—	—	—	—	—	—	—	—
Dairies	—	—	—	—	—	—	—	—	—	—	—	—
Marriage Palaces	6	75	2	25	—	—	—	—	1	100	—	—
Printing Presses	1	100	—	—	—	—	—	—	—	—	—	—
Chemists	1	6.25	2	12.5	2	12.5	3	18.75	2	12.5	—	—

Of the 87 respondents who perceived some health or safety risk, 71 estimated a shortest warning period required to reduce the risk. 30 respondents estimated the period to be twenty minutes 19 estimated the time to be one hour, 8 as 4 hours, 9 as 8 hours, 4 as 24 hours and only 1 estimated the time to be more than 24 hours.

This warning period is required to switch on the generators and for arranging fuel, not ordering or making eaTable s which can get spoilt, making alternate arrangements in cases of interruptions occuring at night (for e.g. storage of drugs etc. in case of chemists) or rescheduling surgical/medical procedures for patients in nursing homes.

WORST TIME FOR AN INTERRUPTION

The next part of the questionnaire deals with establishing the worst time for an interruption to occur. This worst time can vary from organisation to organisation depending on their area of business. Respondents were asked first to identify the worst month for an interruption. They were asked to check box or boxes for a month or months of the year. Apart from the twelve months of the year, a thirteenth choice provided was all months the same' response where the respondent could not identify any month as the worst. Respondents were asked to indicate any number of months, interruptions in which they felt were responsible for more losses than in others.

58.67% of respondents indicated that for their business organisations, all months were the same. The rest 41.33% checked different months indicating some specific months. (Fig. 4.19)

Similarly, information was gathered where respondents indicated the worst day of the week and also worst time during the day. This resulted in identifying the worst time of day, worst day of the week and the worst month of the year for all respondents.

WORST MONTH OF THE YEAR

As indicated in Table 4.11, 176 respondents i.e. 58.67% have selected the choice 'all months the same' meaning that for them

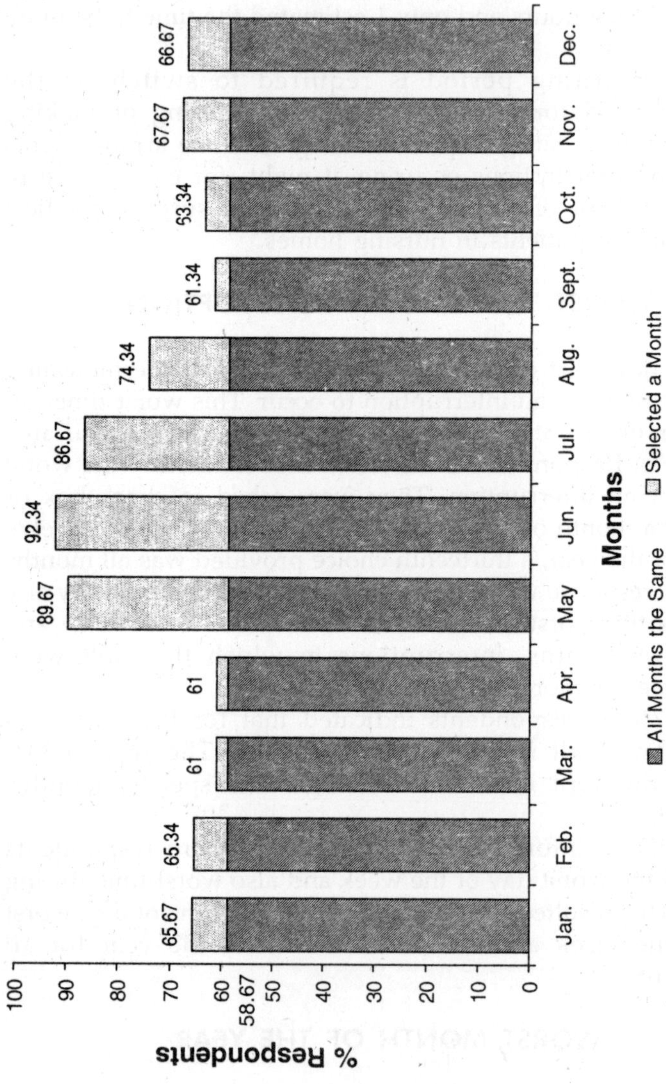

FIG. 4.19
Worst Month

TABLE 4.11
Respondents Selecting Worst Month

Categories	*Selected Particular months*		*All months the same*	
	N	%	*N*	%
Nursing Homes	11	57.89	8	42.11
Cinemas	10	50.00	10	50.00
Restaurants	24	80.00	6	20.00
Shops/Showrooms	13	48.15	14	51.85
Petrol Pumps	4	17.39	19	82.61
Hotels	7	46.67	8	53.33
Diag. Labs.	2	13.33	13	86.67
Motor Garages	2	14.29	12	85.71
Colour Labs.	5	35.71	9	64.29
Dairies	13	68.42	6	31.58
Marriage Palaces	3	20.00	12	80.00
Dry Cleaners	15	100.00	—	—
Comp. Centres	9	52.94	8	47.06
Printing Presses	2	14.29	12	85.71
Chemists	—	—	16	100.00
Cable Operators	—	—	14	100.00
Beauty Parlours	4	30.77	9	69.23
All Respondents	124	41.33	176	58.67

all months are the same and they would suffer losses equally in all the months of the year. 100% of chemists and cable operators have selected 'all months the same as for their area of operations no particular month is worse than the others. 82.61% petrol pumps, 85.71% printing presses and 80% marriage palaces also consider all months to be the same.

41.33% respondents have selected different months which they feel account for more losses as compared to other months of the year.

Of the 124 respondents who have selected particular months to be the worst, the selected months are as shown in Table 4.12. According to it, the maximum number of

TABLE 4.12
Worst Month Selections

Months	*No. of Respondents selecting a month*	*%age*
January	21	16.9
February	20	16.1
March	7	5.6
April	7	5.6
May	93	75
June	101	81.4
July	84	67.7
August	47	37.9
September	8	6.4
October	14	11.2
November	27	21.7
December	24	19.3

* No of respondents total more than 124 as respondents could choose as many months as they wished.

respondents out of these 124, i.e. 81.4% have selected June as the worst month, followed by May by 75% and July by 67.7% respondents.

This shows that due to severity of summers in the area of research, the business operations are affected if there is an electrcity failure since airconditioning becomes important in these months. (Fig. 4.19) In months of March and April and also Sept. and Oct., few respondents have indicated losses as the weather is pleasant. Again November, December, January and February have been chosen as worst months by approximately 20% of respondents as for businesses like drycleaners these are the major earning months.

Worst Day of the Week

Similarly information was collected for the worst day of the week. Respondents were asked to select particular days which are worse than the others or select 'all days the same' choice. A third choice given here was 'all weekdays the same', assuming that for most of the organisations Sunday is not a working day. The results for worst day are shown in Table 4.13.

TABLE 4.13
Respondents Selecting Worst Day

Categories	*Selected Particular Days*		*All days the same*		*All weekdays the same*	
	N	%	N	%	N	%
Nursing Homes	—	—	17	89.47	2	10.53
Cinemas	7	35.00	13	65.00	—	—
Restaurants	10	33.33	18	60.00	2	6.67
Shops/showrooms	7	25.93	6	22.22	14	51.85
Petrol Pumps	2	8.70	19	82.61	2	8.7
Hotels	4	26.67	9	60.00	2	13.33
Diag. Labs.	—	—	12	80.00	3	20.00
Motor Garages	3	21.43	9	64.29	2	14.29
Colour Labs.	1	7.14	5	35.71	8	57.14
Dairies	—	—	19	100.00	—	—
Marriage Palaces	2	13.33	13	86.67	—	—
Dry Cleaners	2	13.33	8	53.33	5	33.33
Comp. Centres	2	11.76	13	76.47	2	11.76
Printing Presses	—	—	6	42.86	8	57.14
Chemists	—	—	16	100.00	—	—
Cable Operators	—	—	14	100.00	—	—
Beauty Parlours	1	7.70	2	15.38	10	76.92
All Respondents	41	13.67	199	66.33	60	20.00

66.33% respondents have not chosen any particular day, but have opted for 'all days the same' choice, indicating that for them all day of the week are equally important. (Fig. 4.20)

100% chemists, cable operators and dairy owners have indicated all days to be the same as chemists and dairies require refrigeration on all days and cable operators' business is also the same for all days. Most of the nursing homes, petrol pumps and marriage palaces have also indicated all days to be equally important.

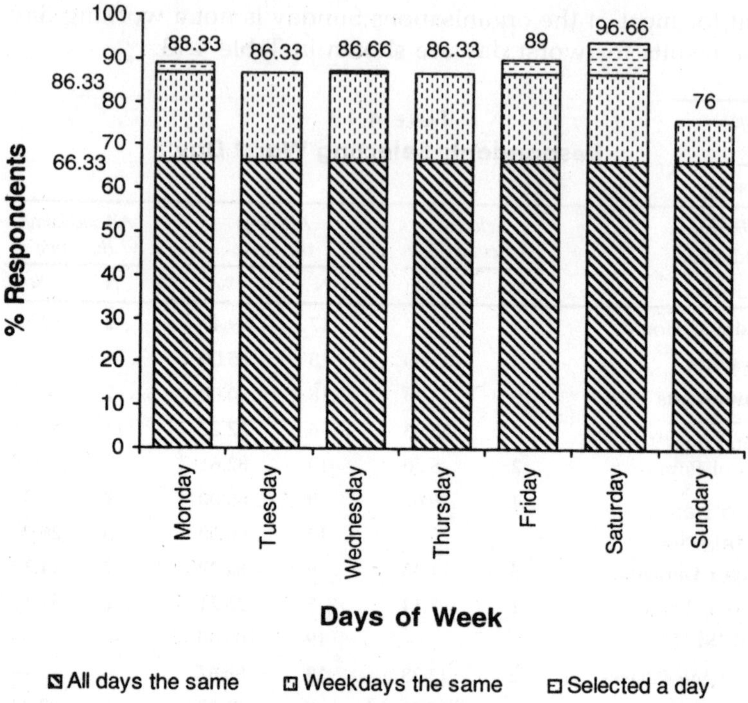

FIG. 4.20
Worst Day

20% respondents have opted for 'all week days the same' choice where they have excluded Sundays and indicated all other days to be the same (Fig. 4.20). These are those organisations which do not operate on Sundays. Printing presses, colour labs and shops/showrooms are the major categories who have opted for this choice.

13.67% respondents have chosen particular days to be the worst for them. Restaurants and cinemas are the major categories who have selected particular days. Table 4.14 shows the worst days for these 13.67% respondents.

As indicated by Table 4.14 out of those having selected a particular worst day, the maximum number of respondents have selected Saturday and Sunday as the worst days for their organisations. As the main categories (as shown in Table 4.13)

TABLE 4.14
Worst Day Selections

Days	*No. of Respondents who selected a day*	*%age*
Monday	6	14.6
Tuesday	—	—
Wednesday	1	2.4
Thursday	—	—
Friday	5	12.1
Saturday	22	53.6
Sunday	29	70.7

*Total doesn't match as any no. of days could be selected.

who have chosen particular days are cinemas, restaurants and hotels, weekend is the worst for them as all these cater to leisure activities.

Worst Time of the Day

Worst time of the day was also selected using the same methodology. Respondents were asked to select any time period/s during the day which was/were worst for their organisations. To assist the respondents, the whole day was divided into 9 time periods. Apart from these time periods, other choices given to respondents were 'all times of night and day the same', 'all working hours the same' and 'worst time is irregular'. Where respondents chose 'all working hours the same', they were asked to indicate their working hours also. So infact these respondents have also chosen certain times of the day to be the worst for them.

44.67% respondents have selected particular times of the day which are the worst for them, while 34.33% have indicated that for them, all times of the day and night are the same (Fig. 4.21).

Another 21% have selected all working hours as the same. (Fig. 4.21) Working hours have been indicated by majority of respondents as 9 a.m. to 9 p.m.

Cable operators have chosen all times to be the same for

TABLE 4.15
Respondents Selecting Worst Time of the Day

Categories	*Selected Particular Days*		*All days the same*		*All weekdays the same*	
	N	%	N	%	N	%
Nursing Homes	2	10.52	15	78.95	2	10.52
Cinemas	14	70	—	—	6	30
Restaurants	24	80	4	13.33	2	6.66
Shops/Showrooms	23	85.19	—	—	4	14.81
Petrol Pumps	2	8.7	12	52.17	9	39.13
Hotels	3	20	12	80	—	—
Diag. Labs.	5	33.33	4	26.67	6	40
Motor Garages	11	78.57	—	—	3	21.43
Colour Labs.	6	42.86	—	—	8	57.14
Dairies	1	5.26	18	94.74	—	—
Marriage Palaces	6	40	9	60.00	—	—
Dry Cleaners	9	60	—	—	6	40
Comp. Centres	8	47.06	—	—	9	52.94
Printing Presses	8	57.14	—	—	6	42.86
Chemists	1	6.25	15	93.75	—	—
Cable Operators	—	—	14	100.00	—	—
Beauty Parlours	11	84.61	—	—	2	15.38
All Respondents	134	44.67	103	34.33	63	21

them. 94.74% of dairies and 93.75% of the chemists have also indicated all times the same for them as refrigeration is required round the clock. Similarly, 78.95% of the nursing homes consider all times the same for them. Of the 44.67% respondents who have selected particular time periods during the day, the results as per Table 4.16 were obtained.

As indicated in Table 4.16, 74.6% respondents feel 5 p.m. to 9 p.m. as the worst time for an interruption to occur, followed by 4 p.m. to 5 p.m. by 58.7% respondents.

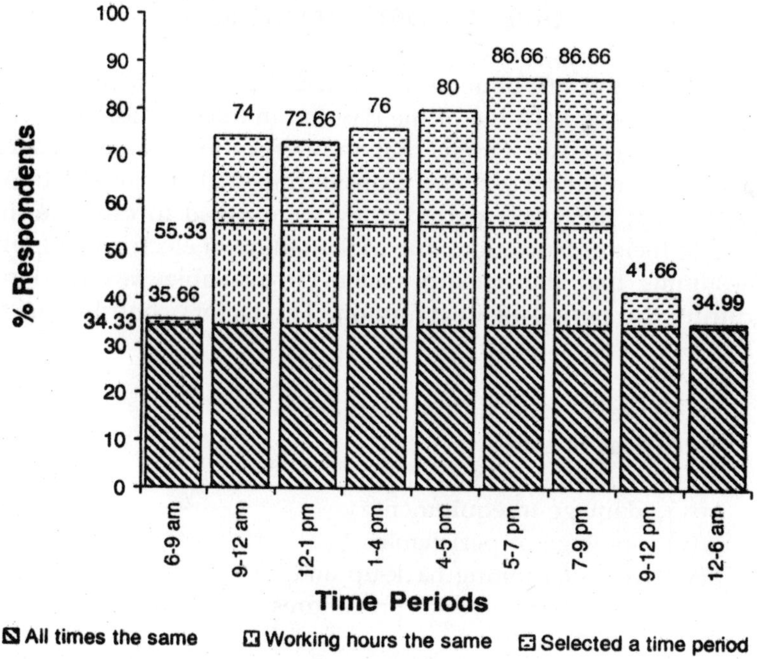

FIG. 4.21
Worst Time of Day

TABLE 4.16
Worst Time Selections

Time Periods	*No. of Respondents*	*%age of Respondents*
6—9 a.m.	4	3.1
9—12 noon	56	44.4
12—1 p.m.	52	41.2
1—4 p.m.	62	49.2
4—5 p.m.	74	58.7
5—7 p.m.	94	74.6
7—9 p.m.	94	74.6
9—12 midnight	22	17.4
12—06 a.m.	2	1.6

Total no. of respondents do not tally as any no. of time periods could be selected by a respondent.

DIRECT COST EVALUATION

After the respondents identified the worst month, worst day and the worst time of the day for an interruption to occur, they were asked to assume that an interruption occured at such a time identified by them. Considering that the interruption occured at the worst period, they were asked to estimate the loss to their organisation, due to the failure of electricity supply assuming that the backup system, if available was working during the period of interruption. The factors contributing to the loss were taken as :

(i) wages paid to idle workers
(ii) loss of sales
(iii) overtime costs
(iv) damage to equipment
(v) spoilage of perishables
(vi) cost of running back-up and
(vii) cost of any special procedures

The time period for which respondents were asked to estimate their losses are :

(i) 2 seconds
(ii) 1 minute
(iii) 20 minutes
(iv) 1 hour
(v) 2 hours
(vi) 4 hours
(vii) 8 hours and
(viii) 24 hours

Wages Paid to Idle Workers

For the factor of wages paid to idle workers no respondent estimated any cost for periods of 2 seconds and 1 minute. It means that if the interruption lasts for only upto 1 minute, their is no loss to any organisation for wages paid to idle workers, though for a period of 20 minutes, as shown by Table 4.17, only two categories of customers estimated losses.

TABLE 4.17
Wages Paid to Idle Workers
(Interruption lasting 20 min.)

Category	*% reporting loss*	*Average loss in Rs.*
Motor garages	14.2	5.36
Printing Presses	14.2	2.86
All respondents	1.33	0.38

These two categories of respondents are motor garages and printing presses. Looking at the whole sample only 1.33% of respondents reported a loss owing to wages paid to idle workers when the interruption lasted 20 minutes. For the total sample the average loss comes out to be Rs. 0.38 though for motor garages and printing presses it is Rs. 5.36 and Rs. 2.86 respectively.

For the same factor, respondents were asked to report losses when the interruption was of one (1) hour. Results obtained are as per Table 4.18.

TABLE 4.18
Wages Paid to Idle Workers
(Interruption lasting one hour)

Category	*% reporting loss*	*Average loss in Rs.*
Restaurants	20.00	22.67
Shops/Showrooms	14.81	18.52
Petrol Pumps	4.35	3.26
Diag. Labs	6.67	6.67
Motor Garages	35.71	42.86
Dry Cleaners	33.33	22.00
Computer Centres	5.88	8.82
Printing Presses	21.43	25.00
All Respondents	8.67	9.28

As compared to 20 minutes, if the interruption is longer, more number and category of respondents have to suffer losses due to wages paid to idle workers. When the period of interruption increases, the reported losses due to wages paid to idle workers also increased, as indicated by Table 4.19.

As the period of interruption increases, the percentage of respondents reporting loss, owing to wages paid to workers who could not work in that period remains consistent at 11% and the categories of business, also remain the same. This indicates that some consumers are dependent on electrical equipment for their main area of business and also that their back-up systems are not able to run all their equipment and their employees have to remain idle. Though there is a sharp increase in this loss upto 8 hours, there is a tapering off at 24 hours. This may be because of the reason that the working day of all these categories of business is seldom more than 8 to 10 hours, so if the interruption is for more than 10-12 hours, the loss due to wages does not increase proportionately.

Loss of Sales

The second contributing factor taken into consideration for losses due to interruptions is 'loss of sales'. Here too, no respondent has reported a loss due to loss of sales for periods of interruption lasting upto 1 minute. For the rest of the interruption durations, the losses are shown in Table 4.20.

For an interruption upto 20 minutes only six categories of respondents have estimated loss of sales. These are restaurants, shops, petrol pumps, chemists, beauty parlours and dairies making a total of 4.67% respondents with an average loss of Rs. 113.20. The percentage as well as categories of respondents rise substantially, as the interruption duration rises upto one hour. 26.67% respondents report loss of sales with almost all categories, barring a few like hotels, marriage palaces and cable operators reporting loss. The average loss for one hour interruption works out to be Rs. 673.23, which denotes a sharp rise from 20 minute average loss of Rs. 113.20.

For an interruption lasting 2 hours, 13.33% of hotels also report loss of sales. In total 34.67% respondents reported loss with an average of Rs. 1061-00, again a substantial rise from one hour average loss. For periods of interruption ranging from 4

TABLE 4.19
Wages Paid to Idle Workers

Category	2 hours		4 hours		8 hours		24 hours	
	%	Avg. loss (Rs.)	%	Avg. loss (Rs.)	%	Avg. loss (Rs.)	%	Avg. loss (Rs.)
Restaurants	20	33.33	20	60.00	20	93.33	20	130.00
Shops/Showrooms	14.81	33.33	14.81	53.70	14.81	85.19	14.81	85.19
Petrol Pumps	4.35	6.52	4.35	9.78	4.35	13.04	4.35	13.04
Diag. Labs	6.67	6.67	6.67	10.00	6.67	13.33	6.67	33.33
Motor Garages	35.71	82.14	35.71	107.14	35.71	139.29	35.71	153.57
Colour Labs	7.14	3.57	7.14	3.57	7.14	3.57	7.14	3.57
Dry Cleaner	46.67	37.33	46.67	61.33	46.67	86.67	46.67	86.67
Computer Centres	5.88	17.65	5.88	35.29	5.88	70.59	5.88	70.59
Printing Presses	50.00	76.43	50.00	115.71	50.00	237.86	50.00	284.29
All Respondents	11.00	17.60	11.00	27.72	11.00	44.77	11.00	52.27

TABLE 4.20
Loss of Sales

Category	20 minute		1 hour		2 hours		4 hours		8 hours		24 hours	
	%	Avg. loss (Rs.)	%	Avg. loss (Rs.)	%	Avg. loss (Rs.)	%	Avg. loss (Rs.)	%	Avg. loss (Rs.)	%	Avg. loss (Rs.)
Nursing Homes	—	—	57.89	1992.11	57.89	2215.79	57.89	3894.74	57.89	4673.68	57.89	5763.16
Cinemas	—	—	20.00	2350.00	20.00	2550.00	20.00	5100.00	20.00	5100.00	20.00	5125.00
Restaurants	6.66	3.33	36.67	716.67	43.33	1153.33	50.00	1560.00	50.00	2383.33	50.00	2550.00
Shops/Showrooms	22.21	1214.81	55.56	3483.33	70.37	6457.41	74.07	9738.89	74.07	13000	74.07	13970
Petrol Pumps	13.05	28.70	26.09	72.61	26.09	156.52	26.09	295.65	26.09	558.70	26.09	891.30
Hotels	—	—	—	—	13.33	100.00	33.33	666.67	33.33	800.00	33.33	1000.00
Diag. Labs.	—	—	33.33	220.00	40.00	350.00	40.00	786.67	40.00	1246.67	40.00	1600.00
Motor Garages	—	—	42.86	153.57	57.14	350.00	57.14	564.29	57.14	885.71	57.14	921.43
Colour Labs.	—	—	28.57	32.14	42.86	107.14	42.86	125.00	42.86	146.43	42.86	153.57
Dairies	5.26	2.63	31.58	36.84	42.11	71.05	42.11	89.47	42.11	107.89	42.11	110.53
Dry Cleaners	—	—	20.00	20.00	46.67	66.67	46.67	120.00	46.67	146.67	46.67	160.00
Computer Centres	—	—	5.88	11.76	5.88	29.41	5.88	58.82	5.88	58.82	5.88	58.82
Printing Presses	—	—	7.14	7.14	35.71	67.86	42.86	192.86	42.86	271.43	42.86	271.43
Chemists	6.25	18.75	18.75	81.25	25.00	175.00	25.00	225.00	25.00	268.75	25.00	318.75
Cable Operators	—	—	—	—	—	—	7.14	7.14	7.14	21.43	7.14	28.57
Beauty Parlours	7.69	3.85	30.77	107.69	30.77	223.08	30.77	430.77	30.77	961.54	30.77	1307.69
All Respondents	4.67	113.20	26.67	673.23	34.67	1061.00	37.33	1768.33	37.33	2258.17	37.33	2506.83

hours to 24 hours the percentage of respondents reporting loss of sales remain consistent at 37.33%, but average loss increases with every increase in duration. The only category which has reported no loss in sales for any duration of interruption is marriage palaces, who have reported a total dependence on back-up systems and all their equipment are able to work with back-up. Similarly cable operators have reported very little loss as compared to other categories as their sale is through monthly subscription and interruptions do not matter as most of subscribers to cable services are regular customers having no other options. The maximum loss in sales has been reported by shops/showrooms with a maximum average being Rs. 12,118.52 as customers either postpone the shopping or find alternatives when there is an electricity failure in the area. Shops are closely followed by nursing homes, restaurants, cinemas which have substantial loss in sales as their major business is largely dependent on electrical equipment. In case of cinemas and restaurants, though basic operations go on with back-up, but the loss of sales in most cases is due to inability of airconditioning system to work on back-up systems and both being leisure activities, customers just do not avail the service in absence of airconditioning in summers which are long and severe in Punjab.

Overtime Costs

The next contributing factor taken into account towards cost to commercial sector is overtime costs. Again for durations of 1 sec. and 1 min. interruption, no loss has been reported by any respondent.

For interruptions lasting from 20 minutes upto 24 hours, losses due to overtime costs have been reported by only three categories of organisations, i.e. cable operators, dry cleaners and printing presses as shown in Table 4.21.

The percentage of respondents indicating a loss owing to overtime costs is very less, a mere 0.33% for 20 minutes, 0.66% for one hour, 1.66% for two hours and for rest of the interruption durations a consistent 2%. The average overtime cost comes out to be between a range of Rs. 1.67 and Rs. 9.00 for durations ranging from 20 minutes to 24 hours. Dry cleaners and printing presses have to complete the undertaken orders on

TABLE 4.21
Overtime Costs

Category	*20 minute*		*1 hour*		*2 hours*		*4 hours*		*8 hours*		*24 hours*	
	%	*Avg. loss (Rs.)*	%	*Avg. loss (Rs.)*	%	*Avg. loss (Rs.)*	%	*Avg. loss (Rs.)*	%	*Avg. loss (Rs.)*	%	*Avg. loss (Rs.)*
Cable Operators	7.14	35.71	7.14	71.43	7.14	107.14	7.14	142.86	7.14	142.86	7.14	142.86
Dry Cleaners	—	—	6.67	13.33	6.67	13.33	13.33	20.00	13.33	20.00	13.33	20.00
Printing Presses	—	—	—	—	21.43	14.29	21.43	17.86	21.43	28.57	21.43	28.57
All Respondents	0.33	1.67	0.66	4.00	1.66	6.33	2.00	8.50	2.00	9.00	2.00	9.00

time and make the workers work overtime to complete the jobs. Cable operators too have to undertake maintenance work after interruptions to check the cable/other equipments.

Damage to Equipment

In contrast to other costs, the cost due to damage to equipment, in case of an electricity interruption occurs even in an interruption of a short duration of 2 seconds.

As indicated in Table 4.22 the percentage of respondents reporting loss due to damage to equipment is quite high at 42.33% at 2 seconds and increasing marginally to 43% for longer durations. The average loss also remains more or less constant with increase in interruption durations ranging from Rs. 631.50 interruption of 2 seconds to Rs. 643.17 interruption for interruption as long as 24 hours. This means that the damage to equipment would occur irrespective of the length of interruption duration as it is caused by a sudden failure in electricity supply. Another significant observation of this loss is that 100% of the categories (all seventeen in the sample) have reported damage to equipment, though the major loss is indicated by nursing homes and cinemas, the average loss for these categories being Rs. 6210.53 and Rs. 1002.50 respectively. These are followed by cable operators, diagnostic labs, hotels and computer centres. The reason may be that these organisations make use of costly electrical equipment which may get damaged due to a sudden failure in electricity supply.

Spoilage of Perishables

Another factor, taken into consideration, which contributes to loss to commercial consumers is spoilage of perishables.

For an interruption of 2 seconds, no respondent has reported any loss, though for a period of 1 min., some loss to perishable has been reported by chemists. When the interruption duration goes up to 20 minutes, two more categories are added, i.e. dairies and colour labs, in total only 1.67% respondents reporting loss with an average loss of Rs. 4.17. For a 1 hour interruption, other consumers like nursing homes, restaurants, some shops, hotels, diag. and colour labs too report spoilage of perishables. In total 10.33% respondents reported spoilage of perishables with an average loss of Rs. 48.33. With every increase in interruption duration

Table 4.22
Damage to Equipment

Categories	2 Second		1 Minute		20 Minutes		1 Hour	
	%	Avg. loss (Rs.)	%	Avg. loss (Rs.)	%	Avg. loss (Rs.)	%	Avg. loss (Rs.)
(1)	(2)	(3)	(4)	(5)	(6)	(7)	(8)	(9)
Nursing Homes	73.68	6078.95	73.68	6078.95	73.68	6078.95	73.68	6078.95
Cinemas	80.00	1002.50	80.00	1002.50	80.00	1002.50	80.00	1002.50
Restaurants	6.67	83.33	6.67	83.33	6.67	83.33	6.67	83.33
Shops/Showrooms	37.04	196.30	37.04	196.30	37.04	196.30	37.04	196.30
Petrol Pumps	30.43	82.61	30.43	82.61	30.43	82.61	30.43	82.61
Hotels	33.33	573.33	33.33	573.33	33.33	573.33	33.33	573.33
Diag. Labs.	66.67	586.67	66.67	586.67	66.67	586.67	66.67	586.67
Motor Garages	28.57	178.57	28.57	178.57	28.57	178.57	28.57	178.57
Colour Labs.	85.71	203.57	85.71	203.57	85.71	203.57	85.71	203.57
Dairies	10.53	5.26	10.53	5.26	10.53	5.26	15.79	57.89
Marriage Palaces	40.00	226.67	40.00	226.67	40.00	226.67	40.00	226.67
Dry Cleaners	6.67	6.67	6.67	6.67	6.67	6.67	6.67	6.67
Computer Centres	70.59	305.88	70.59	305.88	70.59	305.88	70.59	305.88
Printing Presses	42.86	132.14	42.86	132.14	42.86	132.14	42.86	132.14
Chemists	31.25	37.50	31.25	37.50	31.25	37.50	31.25	37.50
Cable Operators	85.71	685.71	85.71	685.71	85.71	685.71	85.71	685.71
Beauty Parlours	23.08	46.15	23.08	46.15	23.08	46.15	23.08	46.15
All Respondents	42.33	631.50	42.33	631.50	42.33	631.50	42.67	634.83

TABLE 4.22 (Contd.)

Categories	2 hours		4 hours		8 hours		24 hours	
	%	Avg. loss (Rs.)	%	Avg. loss (Rs.)	%	Avg. loss (Rs.)	%	Avg. loss (Rs.)
(1)	(10)	(11)	(12)	(13)	(14)	(15)	(16)	(17)
Nursing Homes	78.95	6210.53	78.95	6210.53	78.95	6210.53	78.95	6210.53
Cinemas	80.00	1002.50	80.00	1002.50	80.00	1002.50	80.00	1002.50
Restaurants	6.67	83.33	6.67	83.33	6.67	83.33	6.67	83.33
Shops/Showrooms	37.04	196.30	37.04	196.30	37.04	196.30	37.04	196.30
Petrol Pumps	30.43	82.61	30.43	82.61	30.43	82.61	30.43	82.61
Hotels	33.33	573.33	33.33	573.33	33.33	573.33	33.33	573.33
Diag. Labs.	66.67	586.67	66.67	586.67	66.67	586.67	66.67	586.67
Motor Garages	28.57	178.57	28.57	178.57	28.57	178.57	28.57	178.57
Colour Labs.	85.71	203.57	85.71	203.57	85.71	203.57	85.71	203.57
Dairies	15.79	57.89	15.79	57.89	15.79	57.89	15.79	57.89
Marriage Palaces	40.00	226.67	40.00	226.67	40.00	226.67	40.00	226.67
Dry Cleaners	6.67	6.67	6.67	6.67	6.67	6.67	6.67	6.67
Computer Centres	70.59	305.88	70.59	305.88	70.59	305.88	70.59	305.88
Printing Presses	42.86	132.14	42.86	132.14	42.86	132.14	42.86	132.14
Chemists	31.25	37.50	31.25	37.50	31.25	37.50	31.25	37.50
Cable Operators	85.71	685.71	85.71	685.71	85.71	685.71	85.71	685.71
Beauty Parlours	23.08	46.15	23.08	46.15	23.08	46.15	23.08	46.15
Al Respondents	43.00	643.17	43.00	643.17	43.00	643.17	43.00	643.17

TABLE 4.23
Spoilage of Perishables

Categories	*1 minute*		*20 minute*		*1 hour*		*2 hours*		*4 hours*		*8 hours*		*24 hours*	
	%	*Avg. loss (Rs.)*	%	*Avg. loss (Rs.)*	%	*Avg. loss (Rs.)*	%	*Avg. loss (Rs.)*	%	*Avg. loss (Rs.)*	%	*Avg. loss (Rs.)*	%	*Avg. loss (Rs.)*
Nursing Homes	—	—	—	—	5.26	26.32	31.58	157.89	47.37	510.53	57.89	1284.21	57.89	1342.11
Restaurants	—	—	—	—	30.00	250.00	60.00	880.00	73.33	1835	76.67	2606.67	76.67	2916.67
Shops/Showrooms	—	—	—	—	3.70	37.04	11.11	129.63	11.11	333.33	11.11	814.81	11.11	814.81
Hotels	—	—	—	—	6.67	33.33	33.33	126.67	33.33	1033.33	33.33	246.67	33.33	280.00
Diag. Labs.	—	—	—	—	6.67	33.33	6.67	33.33	6.67	66.67	66.67	6.67	6.67	133.33
Colour Labs.	—	—	7.14	28.57	7.14	28.57	7.14	28.57	7.14	28.57	7.14	28.57	7.14	28.57
Dairies	—	—	10.53	13.16	52.63	60.53	78.95	281.58	84.21	373.68	84.21	557.89	84.21	768.42
Chemists	6.25	6.25	12.50	37.50	43.75	193.75	93.75	687.50	100.00	1712.50	100.00	2875.00	100.00	3343.75
All Respondents	0.33	0.33	1.67	4.17	10.33	48.83	21.33	173.50	24.67	417.17	25.67	621.00	25.67	699.00

the percentage of respondents and also the average loss goes on increasing, it being 21.33% and Rs. 173.50 for 2 hours interruption and 24.67% with an average loss of Rs. 417.17 for an interruption of 4 hours. The main respondents reporting such a loss are those dealing either with eaTable s or drugs/ chemicals which require a certain temperature to be maintained or chemicals getting spoiled due to stoppage of work.

For interruption duration of 8 hours, 25.67% respondents report an average loss of Rs. 621.00, this going up to Rs. 699.00 for a 24 hours interruption. The major loss occurs when the interruption is without warning and occurs at night when the respondent is not present at the place of business to switch on the back-up.

Cost of Running Back-up

Another cost which has to be borne by almost all the respondents is the cost of running the back-up systems, as most back-up systems run on diesel. Table 4.24 indicates the cost of running back-up to different categories of respondents.

For an interruption of 2 seconds, 21.05% of nursing homes indicate a cost for running back-up, the average cost being Rs. 33.16. At 20 minutes, all categories of respondents report costs; 84.33% respondents contributing to an average cost of Rs. 30.74 per 20 minutes. This percentage of respondents remain constant for all durations through out, though the average cost rises consistently, this being Rs. 82.67 for one hour, Rs. 168.20 for two hours, Rs. 327.92 for four hours and Rs. 653.30 and Rs. 1201.65 for interruptions of 8 hours and 24 hours. The major contributors to this cost are nursing homes, cinemas, restaurants, hotels and marriage palaces, probably because of great dependence on electrical equipment and these being big consumers of electricity having relatively more connected load. This is so, as these categories of consumers make use of big generating systems consuming more fuel.

Cost of Special Procedures

A few consumers incur cost for making use of some special procedures. Respondents were asked to indicate the cost of any such special procedures adopted in case of interruptions. These costs are shown in Table 4.25.

Table 4.24
Cost of Running Back-up

Categories	2 Second		1 Minute		20 Minutes		1 Hour	
	%	Avg. loss (Rs.)	%	Avg. loss (Rs.)	%	Avg. loss (Rs.)	%	Avg. loss (Rs.)
(1)	(2)	(3)	(4)	(5)	(6)	(7)	(8)	(9)
Nursing Homes	21.05	33.16	21.05	33.16	94.74	78.89	94.74	183.95
Cinemas	—	—	—	—	100.00	52.75	100.00	156.75
Restaurants	—	—	6.67	0.53	86.67	39.20	86.67	106.83
Shops/Showrooms	—	—	—	—	85.19	19.67	85.19	53.33
Petrol Pumps	—	—	—	—	100.00	16.61	100.00	39.78
Hotels	—	—	—	—	100.00	61.87	100.00	178.00
Diag. Labs.	—	—	—	—	93.33	16.13	93.33	43.47
Motor Garages	—	—	—	—	100.00	33.14	100.00	86.79
Colour Labs.	—	—	—	—	71.43	21.43	71.43	66.43
Dairies	—	—	—	—	21.05	2.42	21.05	4.74
Marriage Palaces	—	—	—	—	100.00	58.93	100.00	170.00
Dry Cleaners	—	—	—	—	73.33	13.93	73.33	35.00
Computer Centres	—	—	—	—	94.12	33.06	94.12	87.35
Printing Presses	—	—	—	—	64.29	20.86	64.29	62.14
Chemists	—	—	—	—	81.25	12.75	81.25	30.94
Cable Operators	—	—	—	—	85.71	20.71	85.71	60.50
Beauty Parlours	—	—	—	—	76.92	12.15	76.92	29.38
All Respondents	1.33	2.10	2.00	2.15	84.33	30.74	84.33	82.67

Table 4.24 (Contd.)

Categories	2 hours		4 hours		8 hours		24 hours	
	%	*Avg. loss (Rs.)*	%	*Avg. loss (Rs.)*	%	*Avg. loss (Rs.)*	%	*Avg. loss (Rs.)*
(1)	(10)	(11)	(12)	(13)	(14)	(15)	(16)	(17)
Nursing Homes	94.74	350.53	94.74	660.00	94.74	294.74	94.74	2087.37
Cinemas	100.00	313.50	100.00	627.00	100.00	1170.00	100.00	1686.75
Restaurants	86.67	223.67	86.67	447.33	86.67	944.67	86.67	1489.37
Shops/Showrooms	85.19	139.44	85.19	213.33	85.19	426.67	85.19	526.00
Petrol Pumps	100.00	79.57	100.00	159.13	100.00	318.26	100.00	845.22
Hotels	100.00	356.00	100.00	712.00	100.00	1424.00	100.00	3586.00
Diag. Labs.	93.33	87.00	93.33	174.40	93.33	348.00	93.33	514.00
Motor Garages	100.00	173.57	100.00	347.14	100.00	694.29	100.00	1057.50
Colour Labs.	71.43	132.86	71.43	265.71	71.43	531.43	71.43	696.43
Dairies	21.05	9.47	21.05	18.95	21.05	37.89	21.05	83.68
Marriage Palaces	100.00	340.00	100.00	680.00	100.00	1360.00	100.00	3144.00
Dry Cleaners	73.33	70.00	73.33	140.00	73.33	274.00	73.33	456.00
Computer Centres	94.12	174.71	94.12	349.41	94.12	698.82	94.12	1036.76
Printing Presses	64.29	124.29	64.29	248.57	64.29	497.14	64.29	1253.57
Chemists	81.25	61.88	81.25	123.75	81.25	247.50	81.25	480.94
Cable Operators	85.71	121.07	85.71	242.14	85.71	484.29	85.71	1487.14
Beauty Parlours	76.92	58.85	76.92	117.69	76.92	235.38	76.92	340.38
All Respondents	84.33	168.20	84.33	327.92	84.33	653.30	84.33	1201.65

TABLE 4.25
Cost of Special Procedures

Category	*20 minute*		*1 hour*		*2 hours*		*4 hours*		*8 hours*		*24 hours*	
	%	*Avg. loss (Rs.)*	%	*Avg. loss (Rs.)*	%	*Avg. loss (Rs.)*	%	*Avg. loss (Rs.)*	%	*Avg. loss (Rs.)*	%	*Avg. loss (Rs.)*
Cinemas	—	—	10.00	150.00	10.00	150.00	10.00	150.00	10.00	150.00	10.00	150.00
Restaurants	3.33	3.33	3.33	3.33	3.33	5.00	3.33	6.67	3.33	6.67	3.33	6.67
Dairies	—	—	15.79	7.37	21.05	16.05	21.05	18.42	21.05	18.42	21.05	18.42
Dry Cleaners	6.67	3.33	6.67	3.33	13.33	5.00	13.33	5.00	13.33	6.67	13.33	6.67
All Respondents	0.67	0.50	2.67	10.97	3.33	11.77	3.33	12.08	3.33	12.17	3.33	12.17

As indicated by the Table, the percentage of respondents adopting any special procedure is quite low, only 3.33% for any interruption from 2 hours to 24 hours. Only a few categories like nursing homes, cinemas, restaurants, dairies and dry cleaners have indicated this cost. The average cost to respondents for any special procedures is a maximum of Rs. 12.17 for an interruption of 24 hours and a minimum of Rs. 0.50 when the interruption is only for 20 minutes. This cost is not an important one, as it affects only a few categories and a small percentage of respondents. Those businesses which do not own back-up systems, have to adopt certain alternative methods (like keeping the perishables in ice) and this cost is associated with such methods only.

Total Costs

All the costs borne by the commercial consumers, are as given on page 87. Apart from these, respondents were asked to indicate any other costs due to power interruptions, but no such cost has been reported. The total cost (a summation of all the costs dealt with till now) has been shown in Table 4.26 and Figure 4.22.

For an interruption lasting 2 seconds, the affected respondents are only 42.67% with an average loss of Rs. 633.60. This loss is caused mainly by damage to equipment. When a sudden interruption occurs, electric equipment gets damaged, resulting in losses. This type of loss is more for business using costly equipment, like nursing homes, cinemas, color labs and cable operators. There is not much variation in the total cost or percentage of respondents reporting losses for a one minute interruption, this being 43.67% with a cost of Rs. 633.99.

There is a major rise in the number of respondents reporting losses when the interruption is for 20 minutes. 90.33% of respondents report an average loss of Rs. 782.11. The increase in cost is not as sharp because apart from damage to equipment and in a few cases spoilage of perishables, the cost of running back-up is not as much in 20 minutes.

For an interruption of one hour duration almost all respondents, i.e. 96.67% report an average loss of Rs. 1463.82. Similarly for two hours interruption 99.67% respondents indicate an average loss of Rs. 2081.57. For all other interruption

TABLE 4.26
Total Costs

Categories	2 Second		1 Minute		20 Minutes		1 Hour	
	%	Avg. loss (Rs.)	%	Avg. loss (Rs.)	%	Avg. loss (Rs.)	%	Avg. loss (Rs.)
(1)	(2)	(3)	(4)	(5)	(6)	(7)	(8)	(9)
Nursing Homes	78.95	6112.11	78.95	6112.11	94.74	6157.75	100.00	8333.95
Cinemas	80.00	1002.50	80.00	1002.50	100.00	1056.75	100.00	3684.25
Restaurants	6.67	83.33	13.33	83.87	96.67	129.37	100.00	829.50
Shops/Shorooms	37.04	196.30	37.04	196.30	92.59	1430.78	100.00	3792.22
Petrol Pumps	30.43	82.61	30.43	82.61	100.00	127.87	100.00	198.26
Hotels	33.33	573.33	33.33	573.33	100.00	635.20	100.00	784.67
Diag. Labs.	66.67	586.67	66.67	586.67	93.33	602.67	100.00	890.13
Motor Garages	28.57	178.57	28.57	178.57	100.00	217.07	100.00	461.79
Colour Labs.	85.71	203.57	85.71	203.57	92.86	253.43	92.86	330.71
Dairies	10.53	5.26	10.53	5.26	42.11	23.42	84.21	167.37
Marriage Palaces	40.00	226.67	40.00	236.67	100.00	285.53	100.00	396.67
Dry Cleaners	6.67	6.67	6.67	6.67	73.33	23.93	93.33	100.33
Computer Centres	30.59	305.88	30.59	305.88	100.00	338.82	100.00	413.82
Printing Presses	42.86	132.14	42.86	132.14	78.57	155.79	78.57	226.43
Chemists	31.25	37.50	37.50	43.75	87.50	106.44	93.75	343.44
Cable Operators	85.71	685.71	85.71	685.71	92.86	742.07	92.86	817.64
Beauty Parlours	23.08	46.15	23.08	46.15	84.62	62.08	100.00	183.23
All Respondents	42.67	633.60	43.67	633.99	90.33	782.11	96.67	1463.82

TABLE 4.26 (Contd.)

Category	2 hours		4 hours		8 hours		24 hours	
	%	Avg. loss (Rs.)	%	Avg. loss (Rs.)	%	Avg. loss (Rs.)	%	Avg. loss (Rs.)
(1)	(10)	(11)	(12)	(13)	(14)	(15)	(16)	(17)
Nursing Homes	100.00	8987.37	100.00	11328.42	100.00	13515.79	100.00	15455.79
Cinemas	100.00	4066.00	100.00	7279.50	100.00	7997.50	100.00	8539.25
Restaurants	100.00	2345.30	100.00	4059.00	100.00	5734.67	100.00	6792.67
Shops/Shorooms	100.00	6956.11	100.00	10535.56	100.00	14522.96	100.00	15593.35
Petrol Pumps	100.00	325.22	100.00	547.17	100.00	972.61	100.00	1832.17
Hotels	100.00	1156.00	100.00	2985.33	100.00	3044.00	100.00	5439.33
Diag. Labs.	100.00	1063.67	100.00	1624.40	100.00	2261.33	100.00	2867.33
Motor Garages	100.00	184.29	100.00	1197.14	100.00	1897.86	100.00	2311.07
Colour Labs.	100.00	475.71	100.00	626.43	100.00	913.57	100.00	1085.71
Dairies	100.00	436.05	100.00	558.42	100.00	780.00	100.00	1038.95
Marriage Palaces	100.00	566.67	100.00	986.67	100.00	1586.67	100.00	3370.67
Dry Cleaners	100.00	199.00	100.00	353.00	100.00	540.67	100.00	736.00
Computer Centres	100.00	527.65	100.00	749.41	100.00	1134.12	100.00	1472.06
Printing Presses	100.00	415.00	100.00	707.14	100.00	1167.14	100.00	1970.00
Chemists	100.00	961.88	100.00	2098.75	100.00	3428.75	100.00	4180.94
Cable Operators	92.86	913.93	100.00	1077.86	100.00	1334.29	100.00	2344.29
Beauty Parlours	100.00	328.08	100.00	594.62	100.00	1243.08	100.00	1694.23
All Respondents	99.67	2081.57	100.00	3204.89	100.00	4241.57	100.00	5124.08

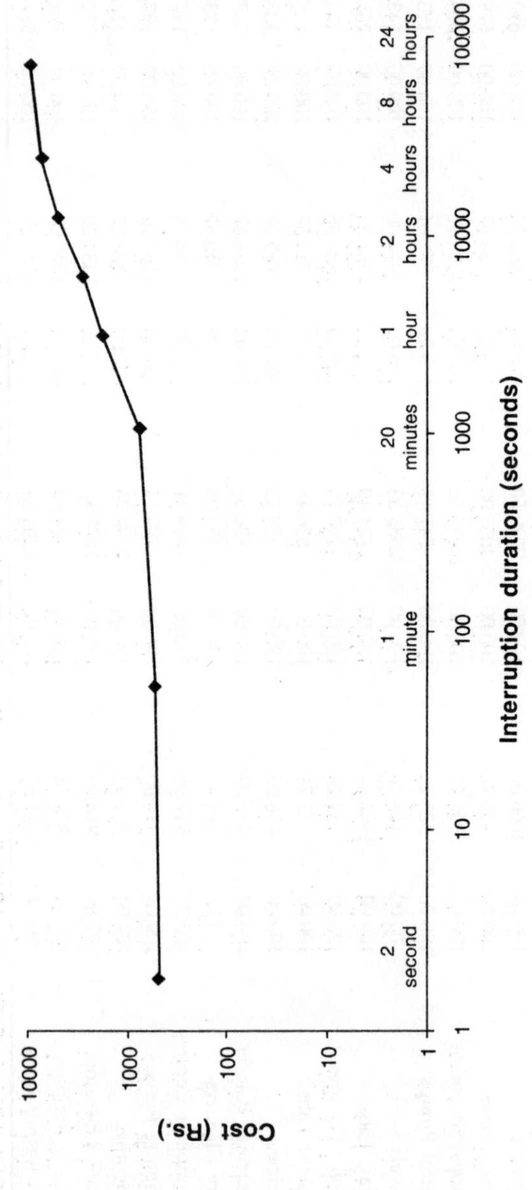

Fig. 4.22
Average Total Cost (Rs.)

durations, i.e. 4 hours, 8 hours and 24 hours. 100% respondents have reported losses. With increase in the duration of interruption average losses also go up, these being Rs. 3204.89, Rs. 4241.57 and Rs. 5124.08.

This clearly indicates that the whole commercial sector in Punjab is affected by electricity interruptions and has been suffering losses, which although varying in magnitude, are distributed throughout the population without any exceptions. The major sufferers are those businesses whose operations depend on electric equipments which is the larger part of the sector. As shown by the Table , the highest sufferers are nursing homes followed by shops. Cinemas and restaurants also suffer largely because of electricity failures.

COST VARIATIONS

After the respondents identified the worst period for an interruption to occur and also estimated losses they would suffer for different durations of interruptions, they were asked to further tell the effect of interruptions on all other months, days and times.

Yearly Variations

Table 4.27 shows the loss for all respondents in different months of the year, relative to that suffered during the worst month.

In previous section, Tables 4.11 and 4.12 indicated that 176 respondents consider all months the same, implying thereby that they would be equally affected during all months. Another 21 respondents indicated January to be their worst month. Apart from those 197, another 24 respondents feel that they would have same loss in Jan. as in their worst month. At the same time rest of the respondents felt their loss would be reduced in this month, 14 (4.67%) saying they would have negligible losses in January as compared to their worst month.

Similarly in the month of February, 221 (73.67%) respondents say they would suffer the most. Most of the rest are able to reduce their cost upto 75% with 2.67% (8) of these estimating it to be negligible in Februrary. Again, in March

TABLE 4.27
Yearly Variation of Interruption Cost

Month	Same as worst Month		10% less		25% less		50% less		75% less		Negligible	
	N	%	N	%	N	%	N	%	N	%	N	%
January	221	73.67	12	4.00	29	9.67	22	7.33	2	0.67	14	4.67
February	221	73.67	12	4.00	32	10.67	23	7.67	4	1.33	8	2.67
March	215	71.67	14	4.67	37	12.33	24	8.00	5	1.67	5	1.67
April	239	79.67	10	3.33	26	8.67	15	5.00	4	1.33	6	2.00
May	281	93.67	1	0.33	3	1.00	4	1.33	2	0.67	9	3.00
June	283	94.34	1	0.33	1	0.33	4	1.33	1	0.33	10	3.33
July	280	93.34	1	0.33	3	1.00	5	1.67	2	0.67	9	3.00
August	275	91.67	2	0.67	5	1.67	7	2.33	1	0.33	10	3.33
September	261	87.00	6	2.00	12	4.00	10	3.33	2	0.67	9	3.00
October	232	77.34	13	4.33	34	11.33	13	4.33	4	1.33	4	1.33
November	231	77.00	11	3.67	25	8.33	14	4.67	3	1.00	16	5.33
December	224	74.67	12	4.00	29	9.67	14	4.67	3	1.00	18	6.00

71.67% would have the maximum loss with only 1.67% estimating it to be negligible (Table 4.27).

From April onwards as the severe Punjab summer sets in more and more respondents indicate their loss as being the maximum, these being 79.67% in April, 93.67% in May, 94.34% and 93.34% in June and July respectively. From August onwards this percentage starts reducing, this being 91.67% in August, 87% in September, and considerably lower at 77.34%, 77% and 74.67% in October, November and December respectively.

This clearly indicates that the summer months, i.e. April through July are the worst months for an interruption to occur for the commercial sector.

Weekly Variations

When asked about the cost variation in different days of the week as compared to the worst day identified by the respondents (Table 4.28) about 92% respondents consider the costs to be the same on Monday, Tuesday, Wednesday and Thursday with the rest indicating some reduction in cost (upto 50%) during these four days. No respondent has said that the cost would be negligible on these days.

On Friday the respondents suffering the maximum loss are 94.67% and the percentage peaks at 97.67% on Saturdays.

There is a decline in this percentage on Sundays when only 73.67% respondents consider the loss to be maximum. 70 respondents (23.33%) estimate their cost to be negligible on a Sunday. Hence the worst day for an interruption to occur would be Saturday closely followed by Friday. Thus it may be concluded that barring on Sundays, interruption in power supply on any week day would lead to considerable loss to a majority of commercial consumers.

Daily Variations

A similar information was collected about the worst time of the day, which is shown in Table 4.29.

When asked about the interruption costs if the interruption occurs between 6 a.m. and 9 a.m., 41% respondents said the cost would be the same as on their worst time. 11% said the cost would be reduced between 10% and 75% and 48% estimated the cost to be negligible.

Table 4.28
Weekly Variation of Interruption Cost

Day of Week	*Same as worst Day*		*10% less*		*25% less*		*50% less*		*75% less*		*Negligible*	
	N	%	N	%	N	%	N	%	N	%	N	%
Monday	275	91.67	6	2.00	11	3.67	8	2.67	–	–	–	–
Tuesday	276	92.00	4	1.33	14	4.67	6	2.00	–	–	–	–
Wednesday	278	92.67	5	1.67	12	4.00	5	1.67	–	–	–	–
Thursday	276	92.00	4	1.33	14	4.33	7	2.33	–	–	–	–
Friday	284	94.67	3	1.00	11	3.67	2	0.67	–	–	–	–
Saturday	293	97.67	1	0.33	5	1.67	1	0.33	–	–	–	–
Sunday	221	73.67	1	0.33	2	0.67	4	1.33	2	0.67	70	23.33

TABLE 4.29
Daily Variation of Interruption Costs

Time Duration	*Same as worst time*		*10% less*		*25% less*		*50% less*		*75% less*		*Negligible*	
	N	%	N	%	N	%	N	%	N	%	N	%
6—9 a.m.	123	41.00	1	0.33	10	3.33	17	5.67	5	1.67	144	48.00
9—12 noon	236	78.00	1	0.33	19	6.33	28	9.33	8	2.67	8	2.67
12—1 p.m.	240	80.00	2	0.67	24	8.00	24	8.00	6	2.00	4	1.33
1—4 p.m.	258	85.33	1	0.33	17	5.67	18	6.00	3	1.00	3	1.00
4—5 p.m.	268	89.33	—	—	10	3.33	16	5.33	4	1.33	2	0.67
5—7 p.m.	279	93.00	—	—	11	3.67	6	2.00	4	1.33	—	—
7—9 p.m.	262	87.33	—	—	11	3.67	10	3.33	—	—	17	5.67
9—12 midnight	163	54.33	3	1.00	12	4.00	10	3.33	4	1.33	108	36.00
12—6 a.m.	112	37.33	1	0.33	7	2.33	13	4.33	6	2.00	161	53.67

When asked about the interruption occuring between 9 a.m. to 12.00 noon, 78% respondents said the cost would be the same as on the worst time with only 2.67% saying it to be negligible. In case of interruption occuring from 12 noon to 1 p.m. 80% respondents said it would be the same as on the worst time. 85.33% respondents estimated the cost to be the maximum if interruption occurs between 1-4 p.m. This percentage goes up to 89.33% if an interruption occurs between 4-5 p.m. The maximum respondents, i.e. 93% said it to be the maximum if an interruption occurs between 5 to 7 p.m. and no respondent has said the cost to be negligible. After 7 p.m. the percentage of respondents reporting maximum losses starts reducing, it being 87.33% if the interruption is from 7 p.m. to 9 p.m., 54.33% if it is from 9 p.m. to 12 midnight and only 37.33% if an interruption occurs from 12 to 6 a.m.

This shows that large number of consumers consider losses to be negligible if the interruption occurs from 12 midnight to 9 a.m. as most of the businesses are closed during this time period.

For those who have said these periods to be the same as the worst time, the reason is that it is not possible for the owners to switch on the generator sets at night and if an interruption occurs it results in spoilage of perishables. (like eatables, medicines etc.)

ROLE OF ADVANCE WARNING IN REDUCING COSTS

After the respondents identified the losses to their organisation due to interruptions, they were asked whether an advance warning would help in reducing their interruption costs. It is important to know whether these costs to the commercial sector consumers can be reduced to some extent by giving them an advance warning about the impending interruption when these are planned ones. Also as to how much minimum warning period is required by different types of respondents as they differ in their operations and factors causing losses. 26.33% respondents replied in the affirmative and said they would be able to reduce the interruption costs if they were given an advance warning. On being asked the minimum warning time required to reduce losses, following results were obtained.

TABLE 4.30
Percentage of Respondents Able to Reduce Cost with Various Advance Warning Times

Advance warning time	*<1 hours*	*1-4 hours*	*5-16 hours*	*17-24 hours*	*1-2 days*	*³3 days*
Cost reduction (%)	6.67	12.00	4.67	1.00	2.00	—

(Percentage shown in this Table are for the total sample of 300 respondents Percentage).

All categories of commercial consumers except beauty parlours have reported reduction in cost, if an advance warning regarding the interruption is provided, though the numbers in each category are few. Maximum percentage is for colour labs. 71.43% respondents from this category reported that they would be able to reduce their interruption costs if they get an advance warning of the interruption. Percentage-wise results for the respondents who are able to reduce their costs are given in Table 4.31.

TABLE 4.31
Category-wise Respondents able to Reduce Interruption Costs with Advance Warning

Category	*% able to reduce cost*
Nursing Homes	36.84
Cinemas	15.00
Restaurants	30.00
Shops/Showrooms	14.81
Petrol Pumps	8.70
Hotels	13.33
Diag. Labs	46.67
Motor Garages	7.14
Colour Labs	71.43
Dairies	42.11
Marriage Palaces	13.33
Dry Cleaners	40.00
Computer Centres	23.53
Printing Presses	42.86
Chemists	37.50
Cable Operators	14.29
Beauty Parlours	—
All Respondents	26.33

When asked if given an advance warning what percentage of cost would they be able to reduce, diagnostic labs estimated an average reduction of 30.67% cost, which is the maximum among all categories. Printing presses, colour labs and dry-cleaners would also be able to reduce costs mostly by rescheduling their work. These percentage reduction in costs is shown in Table 4.32.

TABLE 4.32
Reduction in Cost with Advance Warning

Category	*Avg. reduction in cost (%)*
Nursing Homes	5.79
Cinemas	5.50
Restaurants	19.50
Shops/Showrooms	9.63
Petrol Pumps	8.70
Hotels	2.00
Diag. Labs	30.67
Motor Garages	3.57
Colour Labs	17.50
Dairies	12.63
Marriage Palaces	2.00
Dry Cleaners	16.00
Computer Centres	8.82
Printing Presses	20.00
Chemists	15.94
Cable Operators	8.93
All Respondents	11.23

For all the respondents, able to reduce cost, if an advance warning about the interruption is given the average reduction in total cost comes out to be 11.23%.

ROLE OF INTIMATION OF INTERRUPTION DURATION

A similar question was asked about reduction in cost in case information about the duration of the interruption is given just after the interruption has started. Only 5% of the total respondents said that they would be able to reduce their interruption cost on getting such an information. 95% respondents said that it would not be possible for them to reduce their interruption cost on getting information about the interruption duration once the interruption has already started, as shown by Table 4.33.

TABLE 4.33
Percentage of Respondents Able to Reduce Cost on Intimation of Interruption Duration

Category	*% able to reduce cost*
Nursing Homes	15.79
Restaurants	20.00
Shops/Showrooms	3.70
Hotels	6.67
Motor Garages	7.14
Dairies	10.53
Printing Presses	7.14
All Respondents	5.00

It is possible for these respondents to reduce their cost on being told about interruption duration once the interruption has set in only in case the interruption duration is 20 minutes or more. This is shown in Table 4.34.

TABLE 4.34
Percentage of Respondents Able to Reduce Cost on Intimation of Various Interruption Durations (Once Interruption has started)

Expected duration	*1 min.*	*20 min.*	*1 hour*	*4 hours*	*8 hours*	*24 hours*
Percentage of respondents able to reduce cost	—	0.33	1.33	2.00	1.33	0.33

0.33% respondents would be able to reduce costs in such a case for a 20 minute interruption, 1.33% for a one hour interruption, 2% for 4 hour, 1.33% for 8 hour and 0.33% for a 24 hour interruption.

These 5% respondents were then asked to indicate as to what percentage of costs would be reduced in case intimation of interruption duration is given at the start of the interruption. The average reduction in percentage interruption costs are as shown in Table 4.35.

TABLE 4.35

Reduction in Cost upon Interruption Duration Intimation

Category	*Reduction in average cost (%)*
Nursing Homes	6.58
Restaurants	9.83
Shops/Showroom	1.85
Hotels	1.67
Motor Garages	7.14
Dairies	1.05
Printing Presses	5.36
All Respondents	2.30

For all respondents, 2.30% cost can be reduced on an average in this case. Respondents indicated that costs can be reduced as some jobs can be rescheduled or fuel arranged if interruption duration is known.

DEMOGRAPHIC INFORMATION

After obtaining the information sregarding interruptions and associated costs, respondents were asked to furnish some information on demographic factors which helps in not only categorising the respondents but also in the normalising procedure.

Commercial Space Occupied

Respondents were asked to report about the commercial space occupied by their organisations, the information is shown in Table 4.36.

TABLE 4.36
Commercial Space (Sq.ft.)

Categories	*Commercial Space*
Nursing Homes	7610.53
Cinemas	16050.00
Restaurants	4090.33
Shops/Showroom	1501.48
Petrol Pumps	5843.48
Hotels	13893.33
Diag. Labs	1534.00
Motor Garages	4427.86
Colour Labs	821.43
Dairies	548.95
Marriage Palaces	158666.67
Dry Cleaners	1015.33
Computer Centres	1767.65
Printing Presses	1296.43
Chemists	596.88
Cable Operators	1164.29
Beauty Parlours	1622.31
All Respondents	4756.50

Maximum Power Bill

Information about maximum power bill was also obtained. Respondents were asked to furnish informatin about the amount of their maximum bill in the last one year. The figures shown indicate the bill for two months as P.S.E.B. sends bills to its commercial consumers once in two months. Table 4.37 shows the average of the maximum power bill of all categories.

TABLE 4.37
Maximum Power Bill

Categories	*Max. Power Bill in Rs.*
Nursing Homes	37094.74
Cinema	47000.00
Restaurants	18143.33
Shops/Showrooms	12840.74
Petrol Pumps	5017.39
Hotels	60866.67
Diag. Labs	13653.33
Motor Garages	16035.71
Colour Labs	9642.86
Dairies	960.53
Marriage Palaces	16266.67
Dry Cleaners	2484.00
Computer Centres	8764.71
Printing Presses	6914.29
Chemists	2678.12
Cable Operators	6982.14
Beauty Parlours	5076.92
All Respondents	16285.83

Most of the respondents have indicated the bill to be maximum for summer months, but some like Petrol pumps, motor garages, printing presses etc. have indicated the bill to be almost same for all months.

Connected Load

Another important demographic information is regarding the connected load of the business organisations. Few respondents were ready to show any document regarding the connected load. Thus, this information is based on either what the respondent said or an estimate made on the spot on the basis of the electric equipment being used by the organisation. Table 4.38 gives the average connected loads of all categories.

TABLE 4.38
Average Connected Load (kws.)

Categories	*Connected Load (in kws.)*
Nursing Homes	57.11
Cinemas	104.25
Restaurants	36.33
Shops/Showrooms	30.19
Petrol Pumps	17.35
Hotels	106.67
Diag. Labs	18.00
Motor Garages	33.50
Colour Labs	12.29
Dairies	34.95
Marriage Palaces	77.47
Dry Cleaners	13.80
Computer Centres	23.00
Printing Presses	29.21
Chemists	10.44
Cable Operators	26.00
Beauty Parlours	16.54
All Respondents	38.55

5

Cost Normalisation and CCDF

COST NORMALIZATION

After getting direct cost estimates from all the respondents for durations varying from 2 seconds to 24 hours, their connected loads were obtained. The costs provided by each respondent were normalized with respect to their connected loads to get cost in Rs./kw. The average cost due to interruption to each group of respondents have yielded cost/kw. for the period of interruption.

Wages Paid to Idle Workers

In case of wages paid to idle workers during an electricity interruption the following costs as shown in Table 5.1 were obtained.

As is indicated by Table 5.1 there is no cost to commercial consumers for 2 seconds or 1 min. When the interruption is for 20 minutes, only two categories of respondents motor garages and printing presses are affected. The average normalized cost for this period of interruption is quite low at Rs. 0.03/kw. This cost increases to Rs. 0.91/kw for 1 hour interruption, though no. of categories affected greatly increase. The highest affected category is motor garages with an average cost of Rs. 4.07/kw.,

TABLE 5.1
Wages Paid to Idle Workers

Categories	*20 min.*	*1 hr.*	*2 hr.*	*4 hr.*	*8 hr.*	*24 hr.*
Nursing Homes	—	—	—	—	—	—
Cinemas	—	—	—	—	—	—
Restaurants	—	2.36	3.41	5.61	9.68	12.49
Shops/Showrooms	—	1.44	2.06	3.09	4.69	4.69
Petrol Pumps	—	0.17	0.34	0.51	0.69	0.69
Hotels	—	—	—	—	—	—
Diagnostic Labs	—	3.33	3.33	5.00	6.67	16.67
Motor Garages	0.37	4.07	6.97	9.05	11.19	11.90
Colour Labs	—	—	0.71	0.71	0.71	0.71
Dairies	—	—	—	—	—	—
Marriage Palaces	—	—	—	—	—	—
Dry Cleaners	—	1.92	2.87	4.39	5.56	5.56
Computer Centres	—	0.29	0.59	1.18	2.35	2.35
Printing Presses	0.24	1.33	4.55	5.83	13.01	15.00
Chemists	—	—	—	—	—	—
Cable Operators	—	—	—	—	—	—
Beauty Parlours	—	—	—	—	—	—
All Respondents	0.03	0.91	1.47	2.14	3.35	4.26

followed by diagnostic labs at Rs. 3.33/kw. For an interruption of 2 hours too motor garages remain the most affected with average cost of Rs. 6.97/kw. For the whole sector this cost becomes Rs. 1.47/kw and Rs. 2.14/kw for an interruption of 2 hours and 4 hours respectively. In case of an 8 hour interruption the highest cost is borne by printing presses at Rs. 13.01/kw. This category is followed by motor garages at Rs. 11.19/kw and restaurants at Rs. 9.68/kw. When the interruption lasts for 24 hours the average normalized cost for wages paid to idle workers is Rs. 4.26/kw.

Loss of Sales

The average normalized loss due to loss of sales during an electricity interruption for various categories in commercial sector is as shown in Table 5.2.

TABLE 5.2
Loss of Sales

Categories	*20 min.*	*1 hr.*	*2 hr.*	*4 hr.*	*8 hr.*	*24 hr.*
Nursing Homes	—	42.84	46.40	76.80	97.42	127.20
Cinemas	—	33.77	28.02	52.71	54.38	54.69
Restaurants	278.11	300.16	376.15	522.51	666.29	676.52
Shops/Showrooms	5.31	85.15	191.81	291.61	371.41	403.50
Petrol Pumps	1.38	2.64	6.80	13.42	24.15	34.26
Hotels	—	—	3.86	9.88	12.99	16.32
Diagnostic Labs	—	21.98	31.55	55.88	98.67	112.07
Motor Garages	—	12.66	25.02	42.50	69.71	71.34
Colour Labs	—	5.97	18.06	21.99	26.99	28.78
Dairies	0.44	6.62	13.14	16.70	20.52	21.18
Marriage Palaces	—	—	—	—	—	—
Dry Cleaners	—	1.60	5.41	9.36	11.36	12.58
Computer Centres	—	0.39	0.98	1.96	1.96	1.96
Printing Presses	—	0.71	4.62	10.48	15.41	15.41
Chemists	1.88	8.71	18.34	23.81	28.76	34.23
Cable Operators	—	—	—	0.36	1.07	1.43
Beauty Parlours	0.77	7.58	15.93	30.11	57.14	126.37
All Respondents	28.56	46.16	67.04	98.92	128.56	139.56

For an interruption of 2 seconds and 1 minute there is no loss of sales to any respondent, but for an interruption of 20 minutes the sector suffers a loss of Rs. 28.56/kw. The highest sufferer being restaurants at Rs. 278.11/kw for a 20 minute interruption at the worst time. For an interruption lasting 1 hour the average normalized cost goes up to Rs. 46.16/kw, all categories except hotels, marriage palaces and cable operators reporting a loss of sales. This loss goes up to Rs. 67.04/kw for an interruption which lasts 2 hours. The highest loss remains to restaurants at Rs. 376.15/kw followed by shops/showrooms at Rs. 191.81/kw. Even for this duration marriage palaces and cable operators report no loss of sales. Average normalized loss of sales increases as interruption duration increases, this being Rs. 98.92/kw, Rs. 128.56/kw and Rs. 139.56/kw for interruption durations of 4, 8 and 24 hours respectively. Marriage palaces remain the only business category which does not report any loss of sales due to power interruptions, as no day to day sales are made by these and bookings are made in advance. Even cable operators have no loss of sales as they charge monthly payments. Loss of sales as reported by some of them is mainly due to loss of goodwill on account of interruption in cable signal due to power outages.

Over-time Costs

Respondents were asked to report over-time costs. This cost includes completing any work after normal business hours which could not be done during the normal hours due to power interruptions. The average normalized overtime costs are as indicated by Table 5.3.

TABLE 5.3
Overtime Costs

Categories	*20 min.*	*1 hr.*	*2 hr.*	*4 hr.*	*8 hr.*	*24 hr.*
Dry Cleaners	—	0.78	0.78	1.34	1.34	1.34
Printing Presses	—	—	0.91	1.15	1.79	1.79
Cable Operators	3.57	7.14	10.71	14.29	14.29	14.29
All Respondents	0.17	0.37	0.58	0.79	0.82	0.82

Only three categories of business have reported any overtime costs, these are drycleaners, printing presses and cable operators. For 2 seconds and 1 minute interruptions there is no overtime cost to any category of business and for a 20 minute interruption only cable operators report a cost of Rs. 3.57/kw with the sector average being Rs. 0.17/kw. For a 1 hour interruption this average goes up to Rs. 0.37/kw and Rs. 0.58/kw for a 2 hour interruption. The highest cost is reported by cable operators which is Rs. 10.71/kw for an interruption of 2 hours. The average cost for the whole sector goes up for Rs. 0.79/kw for a 4 hour interruption. For interruptions of 8 hours and 24 hours, this average remains at Rs. 0.82/kw. This cost doesn't play a major role in the total cost of electricity interruptions as only three business categories have reported losses which are quite low as compared to other costs.

Damage to Equipment

Damage to equipment is a factor which causes losses to all categories of business in the sample. It is also a cost which is borne by respondents even if the interruption is for 2 seconds. This cost doesn't increase with an increase in duration of interruption but remains more or less constant for all interruption durations.

TABLE 5.4
Damage to Equipment

Categories	*2 sec.*	*1 min.*	*20 min.*	*1 hr.*	*2 hrs.*	*4 hr.*	*8 hr.*	*24 hr.*
Nursing Homes	75.79	75.79	75.79	75.79	81.06	81.06	81.06	81.06
Cinemas	9.30	9.30	9.30	9.30	9.30	9.30	9.30	9.30
Restaurants	5.86	5.86	5.86	5.86	5.86	5.86	5.86	5.86
Shops/Showroom	12.96	12.96	12.96	12.96	12.96	12.96	12.96	12.96
Petrol Pumps	7.72	7.72	7.72	7.72	7.72	7.72	7.72	7.72
Hotels	12.15	12.15	12.15	12.15	12.15	12.15	12.15	12.15
Diagnostic Labs	21.97	21.97	21.97	21.97	21.97	21.97	21.97	21.97
Motor Garages	6.43	6.43	6.43	6.43	6.43	6.43	6.43	6.43
Colour Labs	19.85	19.85	19.85	19.85	19.85	19.85	19.85	19.85
Dairies	1.10	1.10	1.10	8.62	8.62	8.62	8.62	8.62
Marriage Palaces	3.22	3.22	3.22	3.22	3.22	3.22	3.22	3.22
Dry Cleaners	0.39	0.39	0.39	0.39	0.39	0.39	0.39	0.39
Computer Centre	13.49	13.49	13.49	13.49	13.49	13.49	13.49	13.49
Printing Presses	3.27	3.27	3.27	3.27	3.27	3.27	3.27	3.27
Chemists	3.85	3.85	3.85	3.85	3.85	3.85	3.85	3.85
Cable Operators	33.69	33.69	33.69	33.69	33.69	33.69	33.69	33.64
Beauty Parlours	2.22	2.22	2.22	2.22	2.22	2.22	2.22	2.22
All Respondents	13.74	13.74	13.74	14.21	14.55	14.55	14.55	14.55

The highest cost for damage to equipment due to electricity interruptions has been reported by nursing homes which suffer heavy losses because of very costly and delicate equipment being used which is liable to be damaged by a sudden interruption. Cable operators and diagnostic labs also make use of costly electronic equipment which gets damaged with sudden interruptions. Damage to electric equipment like bulbs and tubes etc. has been reported by almost all the respondents.

Spoilage of Perishables

Spoilage of perishables also poses a great problem for some categories of respondents. Businesses dealing with eatables, medicines, chemicals etc. report spoilage of perishables. Some chemists have reported spoilage even for as small a duration as one minute when the interruption occured at a time when back-up could not be switched on, as some medicines are required to be kept at a certain temperature. Normalized costs associated with spoilage of perishables have been shown in Table 5.5.

TABLE 5.5
Spoilage of Perishables

Categories	*1 min.*	*20 min.*	*1hr.*	*2 hr.*	*4 hr.*	*8 hr.*	*24 hr.*
Nursing Homes	—	—	1.75	6.08	11.62	26.10	26.95
Cinemas	—	—	—	—	23.33	33.33	33.33
Restaurants	—	—	21.70	63.08	96.74	146.49	167.84
Shops/Showroom	—	—	1.85	8.56	19.44	40.74	40.74
Hotels	—	—	0.56	2.23	17.50	4.92	5.29
Diagnostic Labs	—	—	16.67	16.67	33.33	33.33	66.67
Colour Labs	—	1.43	1.43	1.43	1.43	1.43	1.43
Dairies	—	2.76	12.03	49.96	65.58	98.38	124.19
Chemists	0.78	3.19	19.20	68.86	173.54	300.02	342.85
All Respondents	0.04	0.41	5.16	15.31	29.73	46.40	54.20

As the interruption duration reaches one hour, no. of categories affected and also the average cost rises sharply. It is Rs. 0.41/kw for 20 minutes and goes up to Rs. 5.16/kw for a one hour interruption. Shops/showrooms indicating spoilage of perishables are those which deal in eatables like icecream parlours, bakeries and sweet shops. For a two hour interruption, average cost is Rs. 15.31/kw with chemists suffering the most at Rs. 68.86/kw followed by restaurants and dairies at Rs. 63.08/kw and Rs. 49.96/kw respectively. This trend continues, there being an increase in average normalized cost at every increase in interruption duration with a maximum of Rs. 54.20/kw for an interruption of 24 hours.

Cost of Running Back-up

Apart from all the costs discussed above, a major cost to all the respondents during an interruption is the cost of running the back-up systems. A majority of respondents have been unable to estimate the cost of running a battery system or even including the initial investment or periodic maintenence costs. The costs shown in Table 5.6 are mainly the costs associated with fuel consumption in engine generating systems which is the major component of running a back-up system.

TABLE 5.6
Cost of Tunning Back-up

Categories	*2 sec.*	*1 min.*	*20 min.*	*1 hr.*	*2 hr.*	*4 hr.*	*8 hr.*	*24 hr.*
Nursing Homes	0.39	0.39	1.31	3.30	6.40	12.34	24.39	44.32
Cinemas	—	—	0.52	1.48	2.96	5.92	10.99	16.24
Restaurants	—	0.06	1.86	4.87	10.99	21.98	50.21	79.20
Shops/Showrooms	—	—	0.85	2.19	8.03	8.78	17.55	21.55
Petrol Pumps	—	—	1.19	2.78	5.56	11.11	22.23	58.30
Hotels	—	—	0.88	2.63	5.27	10.54	21.07	55.66
Diagnostic Labs	—	—	0.99	2.49	4.99	10.00	19.95	27.13
Motor Garages	—	—	1.27	3.16	6.32	12.65	25.30	36.92
Colour Labs	—	—	1.40	4.11	8.23	16.45	32.91	44.45
Dairies	—	—	0.42	0.82	1.64	3.27	6.54	15.39
Marriage Palaces	—	—	0.81	2.34	4.68	9.37	18.73	42.31
Dry Cleaners	—	—	0.98	2.39	4.78	9.56	18.52	29.09
Computer Centres	—	—	1.60	4.20	8.39	16.78	33.56	46.39
Printing Presses	—	—	0.88	2.71	5.43	10.86	21.71	58.92
Chemists	—	—	1.29	3.10	6.19	12.38	24.77	48.34
Cable Operators	—	—	0.86	2.55	5.12	10.23	20.47	61.55
Beauty Parlours	—	—	0.71	1.62	3.25	6.49	12.98	17.49
All Respondents	0.02	0.03	1.08	2.81	6.06	11.43	23.38	42.29

For an interruption of 2 seconds, cost has been reported only by nursing homes, some of which have automatic systems which switch on immediately at even a momentary interruption and cost has to be incurred for this small a period too. The overall sector cost for 2 seconds is Rs. 0.02/kw and Rs. 0.03/kw for a 1 minute interruption with only restaurants reporting a cost apart from nursing homes.

For an interruption which lasts 20 minutes all categories of respondents have indicated costs, with restaurants spending Rs. 1.86/kw, mainly because of their dependence on airconditioning. Similarly computer centres too estimate a cost of Rs. 1.60/kw.

For a one hour interruption, average for the sector rises to Rs. 2.81/kw, maximum being spent by restaurants, i.e. Rs. 4.87/kw followed again by computer centres at Rs. 4.20/kw and nursing homes at Rs. 3.30/kw.

For a two hours interruption, the average cost is Rs. 6.06/kw. This cost increases proportionally with every increase in interruption duration with a maximum of Rs. 42.29/kw for a period of 24 hours. When estimating costs for 24 hours, most of the respondents have reported cost for a period which is the working hours of the organisation as that is the maximum when the back-up would work even though the interruption may be for a longer duration.

Cost of Special Procedures

Many organisations make use of certain alternative/special procedures to cut losses. The detailed report on these costs is shown in Table 5.7.

TABLE 5.7
Cost of Special Procedures

Categories	*20 min.*	*1 hr.*	*2 hr.*	*4 hr.*	*8 hr.*	*24 hr.*
Nursing Homes	—	0.58	0.58	0.58	0.58	0.58
Cinemas	—	1.20	1.20	1.20	1.20	1.20
Restaurants	0.33	0.33	0.50	0.67	0.67	0.67
Dairies	—	1.24	2.63	3.06	3.06	3.06
Dry Cleaners	0.37	0.37	0.51	0.51	0.65	0.65
All Respondents	0.05	0.25	0.36	0.40	0.41	0.41

For an interruption of 2 seconds or 1 minute, no organisation used any special procedure. For a 20 minutes interruption, restaurants and dry cleaners have reported certain costs which are Rs. 0.33/kw and Rs. 0.37/kw respectively. For the whole sector this cost is Rs. 0.05/kw. For a one hour interruption this cost is Rs. 0.25/kw and Rs. 0.36/kw for two hours; the maximum being Rs. 0.41/kw for an 8 hour or 24 hours interruption. This cost is relatively less as most of the respondents who use any special procedures have indicated using ice instead of refrigeration to save perishables from spoilage or in some cases work is done manually instead of machines resulting in some costs.

Total Costs

All these normalized costs which commercial consumers have to face and which have been dealt individually, make up total cost figures which are the total normalized costs of electricity interruptions.

TABLE 5.8
Total Cost

Categories	*2 sec.*	*1 min.*	*20 min.*	*1 hr.*	*2 hr.*	*4 hr.*	*8 hr.*	*24 hr.*
Nursing Homes	76.18	76.18	77.10	124.27	140.52	182.40	229.55	280.11
Cinemas	9.30	9.30	9.82	35.75	41.48	92.46	109.20	114.77
Restaurants	5.86	5.92	286.17	335.29	459.99	653.37	879.19	942.58
Shops/Showrooms	12.96	12.96	19.12	103.59	223.50	335.88	447.36	483.44
Petrol Pumps	7.72	7.72	10.28	13.31	20.42	32.76	54.78	100.96
Hotels	12.15	12.15	13.03	15.34	23.50	50.06	51.14	89.43
Diagnostic Labs	21.97	21.97	22.96	66.45	78.51	126.19	180.59	244.51
Motor Garages	6.43	6.43	8.07	26.32	44.74	70.63	112.63	126.59
Colour Labs	19.85	19.85	22.66	31.36	48.28	60.44	81.89	95.22
Dairies	1.10	1.10	4.71	29.32	75.98	97.22	137.11	172.44
Marriage Palaces	3.22	3.22	4.03	5.56	7.91	12.59	21.96	45.53
Dry Cleaners	0.39	0.39	1.74	7.46	14.74	25.56	37.81	49.60
Computer Centres	13.49	13.49	15.09	18.37	23.45	33.41	51.37	64.20
Printing Presses	3.27	3.27	4.39	8.04	18.78	31.58	55.19	94.40
Chemists	3.85	4.63	10.20	34.86	97.24	213.59	357.40	429.26
Cable Operators	33.69	33.69	38.11	43.38	49.52	58.56	69.51	110.95
Beauty Parlours	2.22	2.22	3.69	11.42	21.40	38.82	72.34	146.08
All Respondents	13.76	13.81	44.03	69.88	105.36	157.96	217.46	256.07

For an interruption lasting 2 seconds the total cost is Rs. 13.76/kw, the major contributor to this cost being nursing homes, which report damage to equipment as a major cost for even this small an interruption. In-fact this cost is mainly because of damage to eqipment when an interruption occurs without the equipment being switched off. The cost remains somewhat the same at Rs. 13.81/kw for a one minute interruption.

The total cost rises to Rs. 44.03/kw when the interruption is of 20 minutes duration. This rise is mainly because of running of generators. Again the cost rises to Rs. 69.88/kw for a one hour interruption, the main contributors being restaurants and nursing homes having average cost of Rs. 335.29/kw and Rs. 124.27/kw respectively. The cost goes on increasing with every increase in interruption duration, this being Rs. 105.36/kw for two hours interruption and Rs. 157.96/kw for every four hour interruption. For an eight hour and 24 hour interruption this cost becomes Rs. 217.46/kw and Rs. 256.07/kw respectively. The major loss bearers for all interruption durations remain restaurants, cinemas, nursing homes, hotels, shops/showrooms, chemists and beauty parlours.

COMMERCIAL COST DAMAGE FUNCTION

The most direct and usable form of customer perceived interruption costs is the customer damage functions (CDF) (R. Billinton, M. Pandey).[1] The CDF indicates the costs associated with power interruption as a function of outage duration. The CDF can be determined for a particular customer class within a sector or for an overall customer sector.

For the purpose of the present study, the CDF is calculated for all the customer classes included in the study and also for the whole sector, which would be called commercial cost damage function (CCDF).

Cost Damage Function—Nursing Homes

Table 5.9 shows the cost damage function for nursing homes for interruptions in electricity ranging from one minute to eight hours. As the table shows even for an interruption as small as one minute the damage function is quite high at

TABLE 5.9
Cost Damage Function—Nursing Homes

Interruption Duration	*CDF*
1 min.	107.03
20 min.	107.83
1 hr.	145.94
2 hr.	157.38
4 hr.	198.38
8 hr.	236.68

Rs. 107.03/kw. As the interruption duration increases, the damage function also increases, showing that longer the interruption duration, higher is the damage function for this class of consumers (Fig. 5.1).

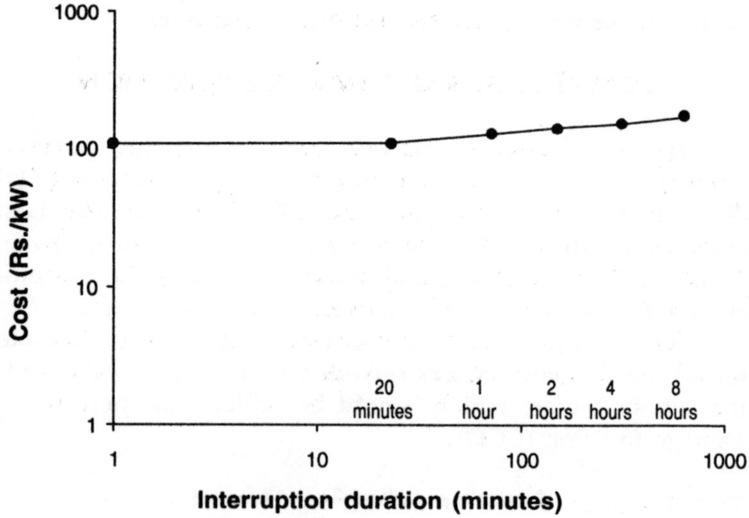

FIG. 5.1
Cost Damage Function—Nursing Homes

Cost Damage Function—Cinemas

The damage function for cinemas is much lower than that of nursing homes. For an interruption of 1 min. the damage function is Rs. 9.28/kw, which increases only marginally to Rs. 9.78/kw. when the interruption duration goes up to 20 minutes. After that it increases considerably for an interruption of one hour and goes on increasing for every increase in interruption duration (Fig. 5.2).

TABLE 5.10
Cost Damage Function—Cinemas

Interruption Duration	*CDF*
1 min.	9.28
20 min.	9.78
1 hr.	34.11
2 hr.	37.65
4 hr.	67.40
8 hr.	74.05

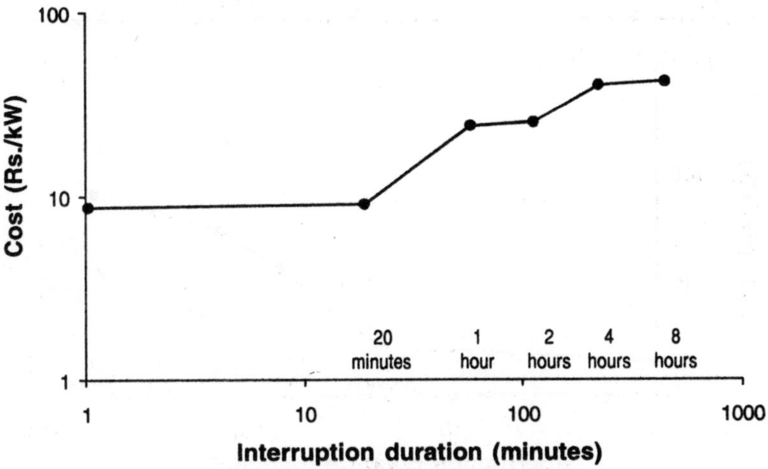

FIG. 5.2
Cost Damage Function—Cinemas

Cost Damage Function—Restaurants

For restaurants, sweet shops or bakeries, selling eatables, initially the damage function is quite low at Rs. 2.28/kw for an interruption of one minute, increasing slightly to Rs. 3.52/kw for an interruption of 20 minutes.

The damage function then increases considerably as the interruption duration increases, as it leads to spoilage of eatables (Fig. 5.3). This indicates that this class of customers is not affected much when the interruption durations are shorter unlike nursing homes where even a momentary interruption can cause great losses.

TABLE 5.11
Cost Damage Function—Restaurants

Interruption Duration	*CDF*
1 min.	2.28
20 min.	3.52
1 hr.	22.58
2 hr.	63.85
4 hr.	110.50
8 hr.	156.12

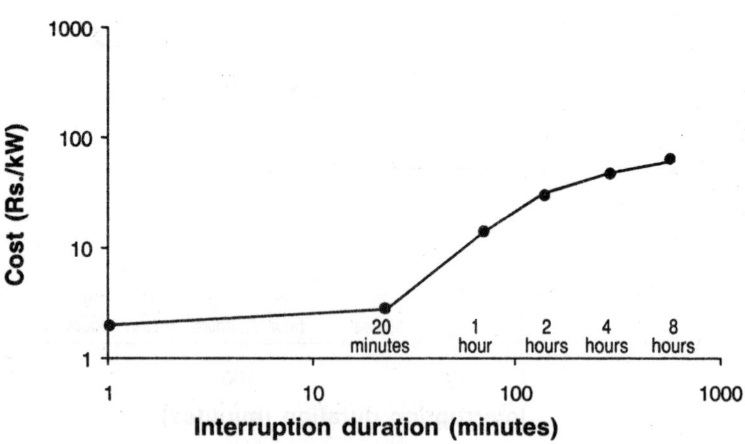

FIG. 5.3
Cost Damage Function—Restaurants

Cost Damage Function—Shops/Showrooms

Table 5.12 shows the cost damage function for shops/showrooms. Here also there is a considerable increase in the damage function for every increase in interruption duration (Fig. 5.4). This is because many of the shops/showrooms, rely on air-conditioning during the summer months, which have been indicated as the worst months by these consumers. Apart from other losses, loss of sales contribute largely to the damage function, as interruptions in electricity have a major impact on sales.

TABLE 5.12
Cost Damage Function—Shops/Showrooms

Interruption Duration	*CDF*
1 min.	6.48
20 min.	47.23
1 hr.	125.17
2 hr.	229.60
4 hr.	347.75
8 hr.	479.36

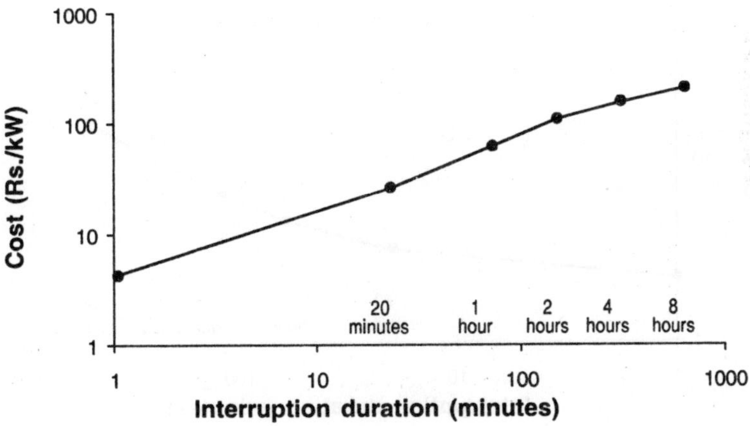

FIG. 5.4
Cost Damage Function—Shops and Showrooms

Cost Damage Function—Petrol Pumps

Table 5.13 shows the damage function for petrol pumps. The cost damage function for this category of consumers is quite low at Rs. 4.76/kw for a 1 minute interruption, going up to Rs. 56.07/kw for an 8 hour interruption (Fig. 5.5). The reason may be that in this class of business, there is no loss of perishables, and not even much loss of sales, as sales continue either with the help of a back-up system or by operating the pumps manually. This class of business is thus affected mildly by electricity interruptions.

TABLE 5.13
Cost Damage Function—Petrol Pumps

Interruption Duration	*CDF*
1 min.	4.76
20 min.	7.37
1 hr.	11.43
2 hr.	18.75
4 hr.	31.54
8 hr.	56.07

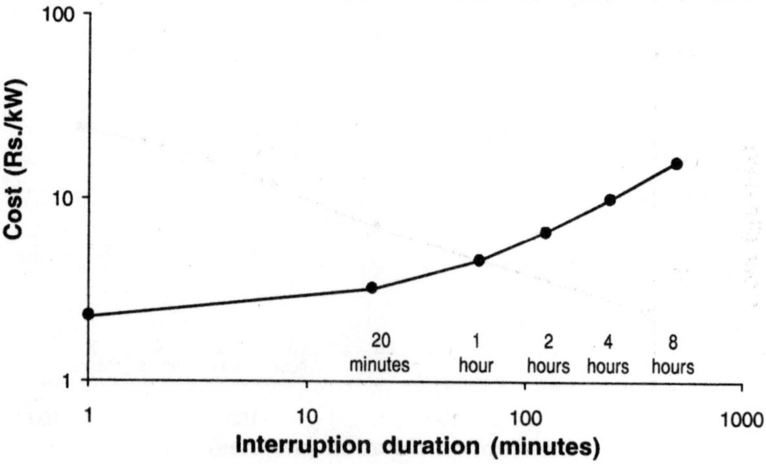

FIG. 5.5
Cost Damage Function—Petrol Pumps

Cost Damage Function—Hotels

For the hotel industry the cost damage function is not very high, starting with Rs. 5.38/kw for an interruption of 1 minute to Rs. 28.54/kw for an 8 hour interruption (Fig. 5.6).

The main reason for this is that most of the hotels are equipped with appropriate back-up systems which can operate all types of equipments. The only cost in that case is the cost of running the back-up. Most of hotels, though bear an initial high cost of having a high capacity generator system. This sector is not affected as much by electricity interruptions because of their full-fledged preparation to face the interruptions.

TABLE 5.14

Cost Damage Function—Hotels

Interruption Duration	*CDF*
1 min.	5.38
20 min.	5.96
1 hr.	7.36
2 hr.	10.84
4 hr.	27.99
8 hr.	28.54

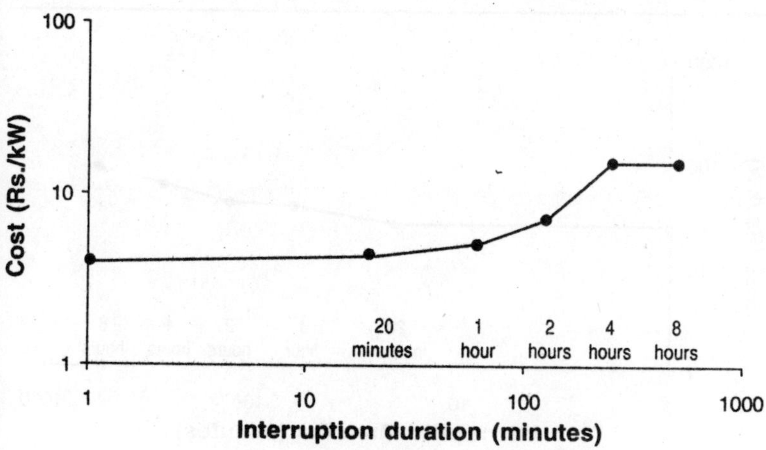

FIG. 5.6

Cost Damage Function—Hotels

Cost Damage Function—Diagnostic Labs

Diagnostic labs which report losses due to damage to equipment, spoilage of perishables and also cost of running back up, have a considerable CDF even for an interruption of one minute.

Organisations such as diagnostic labs, which use sophisticated electric/electronic gadgets have a higher CDF even on momentary interruptions, which doesn't increase much when the duration of interruption increases upto 20 minutes. Though after that the damage function starts to rise proportionately with the duration of interruption as costs due to spoilage of perishables and running of back-up systems start contributing to the damage function (Fig. 5.7).

TABLE 5.15
Cost Damage Function—Diagnostic Labs

Interruption Duration	*CDF*
1 min.	32.59
20 min.	33.48
1 hr.	49.45
2 hr.	59.09
4 hr.	90.24
8 hr.	125.63

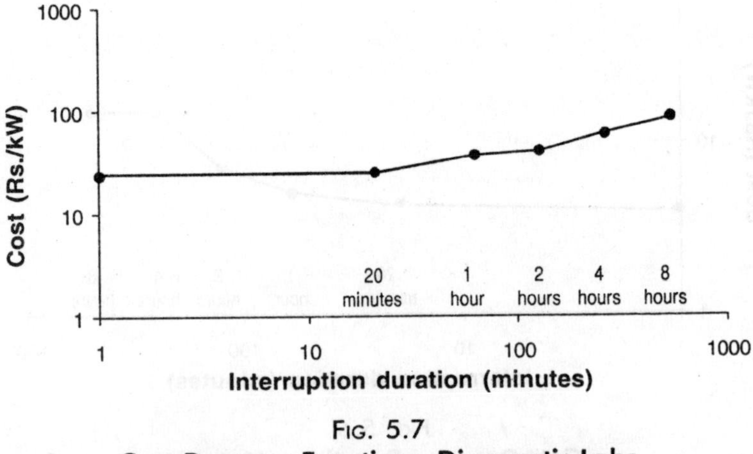

FIG. 5.7
Cost Damage Function—Diagnostic Labs

Cost Damage Function—Motor Garages

Table 5.16 shows the cost damage function associated with motor garages, which deal with servicing of automobiles. The overall CDF for motor garages is quite low, though it is considerable at Rs. 5.33/kw for an interruption of 1 minute. The CDF reaches at Rs. 56.65/kw for an interruption of 8 hours, which is almost the whole working day (Fig. 5.8). The major contribution towards CDF of this category is costs associated with loss of sales and wages paid to idle workers as many jobs cannot be undertaken during electricity outages.

TABLE 5.16
Cost Damage Function—Motor Garages

Interruption Duration	*CDF*
1 min.	5.33
20 min.	6.48
1 hr.	13.78
2 hr.	23.41
4 hr.	35.74
8 hr.	56.65

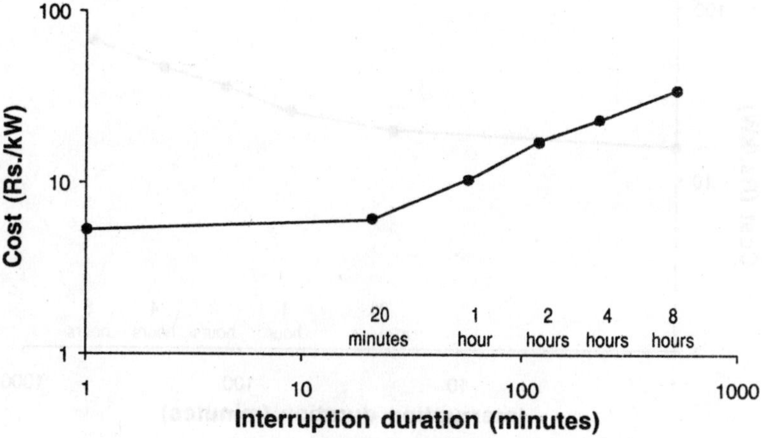

FIG. 5.8
Cost Damage Function—Motor Garages

Cost Damage Function—Colour Labs

Table 5.17, shows the cost damage function of colour labs, which deal with photography, video movie making and developing/printing of photo films. The CDF for 1 min. interruption for this category is Rs. 16.57 which goes up to Rs. 20.63 for 20 minutes. There is a continuous increase in the damage function with every increase in the interruption duration, the maximum being Rs. 74.36 for an interruption of 8 hours (Fig. 5.9). This cost is an amalgamation of costs related to damage to equipment and perishables, wages paid to idle workers and also loss of sales.

TABLE 5.17
Cost Damage Function—Colour Labs

Interruption Duration	*CDF*
1 min.	16.57
20 min.	20.63
1 hr.	26.92
2 hr.	38.72
4 hr.	50.99
8 hr.	74.36

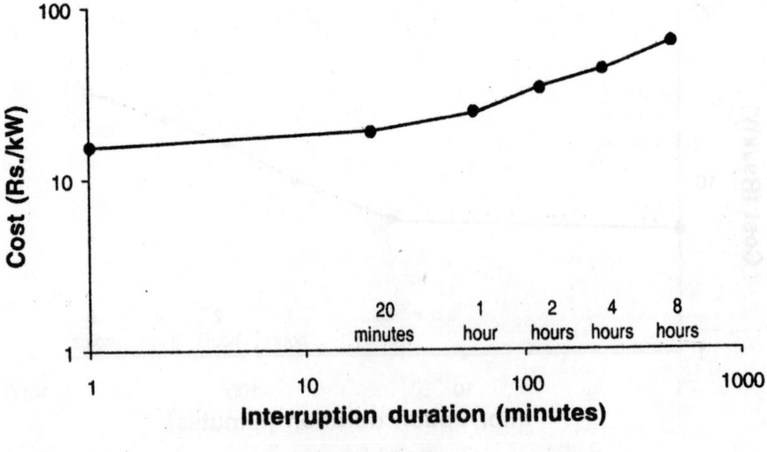

FIG. 5.9
Cost Damage Function—Colour Labs

Cost Damage Function—Dairies

Table 5.18 shows the cost damage function for dairies, which have minimal costs associated with electricity interruptions for interruption durations of 1 min. and 20 minutes, those being Rs. 0.15 and Rs. 0.67 respectively. Though

TABLE 5.18
Cost Damage Function (Dairies)

Interruption Duration	*CDF*
1 min.	0.15
20 min.	0.67
1 hr.	4.79
2 hr.	12.48
4 hr.	15.98
8 hr.	22.32

as the interruption duration increases to more than 1 hour, the CDF increases substantially, which is mainly because of spoilage of dairy products which need refrigeration, more so in summer months. For an interruption of 8 hours the CDF, goes up to Rs. 22.32/kw. (Fig. 5.10). Most of the dairies take up some

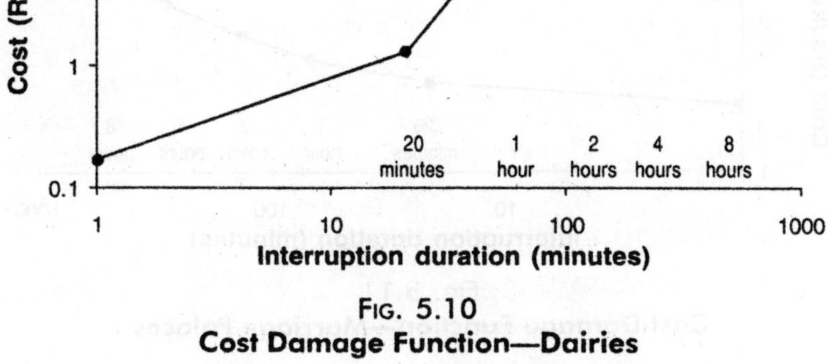

FIG. 5.10
Cost Damage Function—Dairies

alternative arrangements like keeping the perishables in ice as these also do not have appropriate generator sets. The major part of the cost is though associated with spoilage of perishables.

Cost Damage Function—Marriage Palaces

Table 5.19, shows the CDF for marriage palaces. The only factor contributing to the damage function for this category is the cost of running the back up systems. Most of the organisations in this category reported of having back-up system which can run all their equipment. The CDF for 1 min. is 2.93/kw, Rs. 3.69/kw for 20 minutes with a maximum being Rs. 20.48/kw. for an interruption duration of 8 hours (Fig. 5.11).

TABLE 5.19
Cost Damage Function—Marriage Palaces

Interruption Duration	*CDF*
1 min.	2.93
20 min.	3.69
1 hr.	5.12
2 hr.	7.31
4 hr.	11.70
8 hr.	20.48

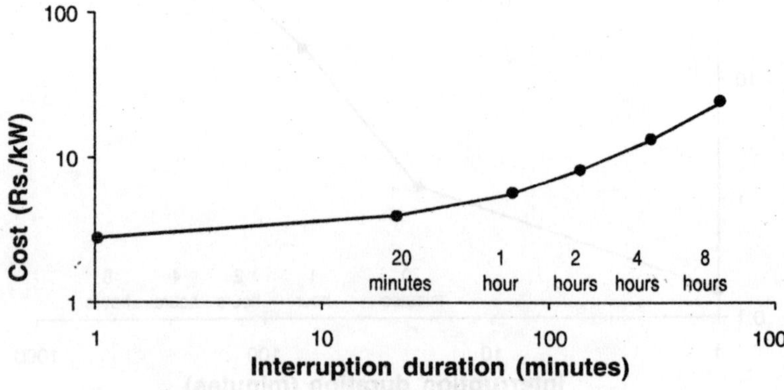

FIG. 5.11
Cost Damage Function—Marriage Palaces

Cost Damage Function—Dry Cleaners

Cost damage function for drycleaners is quite less for an interruption of one minute, i.e., Rs. 0.48/kw. This goes up to Rs. 1.73/kw. for a 20 minute interruption. As the interruption duration increases, the CDF increases quite a great deal, to Rs. 7.27/kw for an interruption of one hour.

TABLE 5.20
Cost Damage Function—Dry Cleaners

Interruption Duration	*CDF*
1 min.	0.48
20 min.	1.73
1 hr.	7.27
2 hr.	14.42
4 hr.	25.58
8 hr.	39.18

The costs go up to Rs. 39.18/kw for an 8 hour interruption (Fig. 5.12). Therefore CDF is high for longer power interruptions.

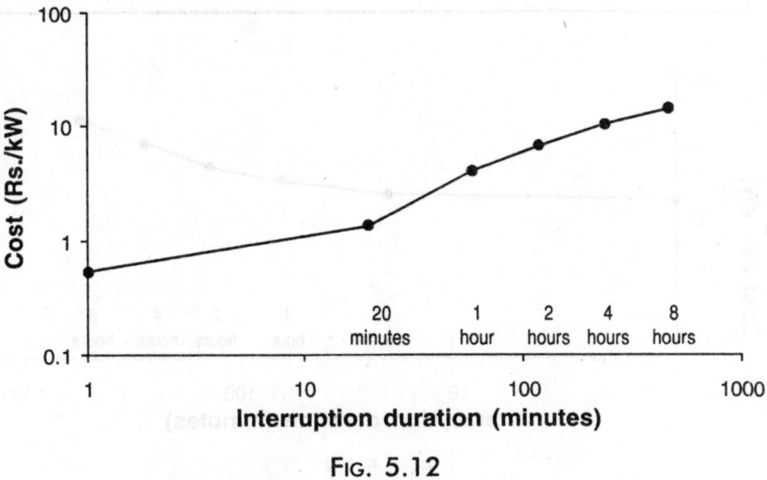

FIG. 5.12
Cost Damage Function—Dry Cleaners

Cost Damage Function—Computer Centres

Table 5.21 shows the CDF for computer centres. In this category organisations dealing with computer education and also commercial computer typing/printing etc. have been taken.

The major cost factor for this category is damage to equipment and also loss of sales and cost of running back-up. For an interruption of one minute the CDF for this sector is Rs. 13.30/kw, and rises with every increase in interruption duration. For an interruption of 8 hours, the CDF goes up to Rs. 49.31/kw. (Fig. 5.13). In computer centres apart from computers which is the main apparatus running on electricity, air-conditioning is of utmost importance.

TABLE 5.21
Cost Damage Function—Computer Centres

Interruption Duration	*CDF*
1 min.	13.30
20 min.	14.73
1 hr.	17.99
2 hr.	22.94
4 hr.	32.58
8 hr.	49.31

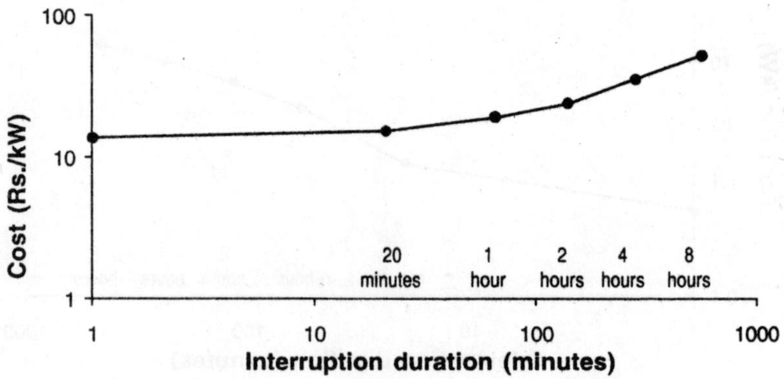

FIG. 5.13
Cost Damage Function—Computer Centres

Cost Damage Function—Printing Presses

Table 5.22 shows the cost damage function for printing presses. This category of organisations suffers mostly on account of wages paid to idle workers, loss of sales and also cost of running back-up systems. For an interruption of one minute the CDF is Rs. 4.52/kw which rises marginally to Rs. 5.33/kw for an interruption of 20 minutes. As the interruption duration increases the CDF also increases proportionately going up to Rs. 39.95/kw for an interruption of 8 hours (Fig. 5.14).

TABLE 5.22
Cost Damage Function—Printing Presses

Interruption Duration	*CDF*
1 min.	4.52
20 min.	5.33
1 hr.	7.75
2 hr.	14.21
4 hr.	24.21
8 hr.	39.95

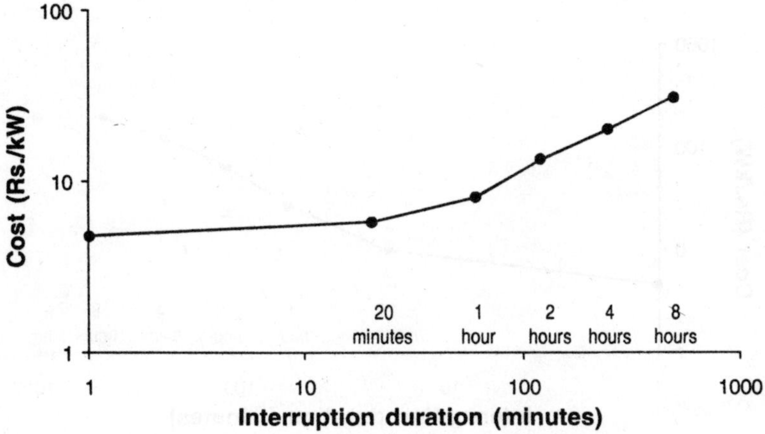

FIG. 5.14
Cost Damage Function—Printing Presses

Cost Damage Function—Chemists

Table 5.23 shows the CDF for chemists—organisations selling medicines/drugs.

The major factor contributing to CDF for this category is spoilage of perishables. Since many medicines/vaccines etc. have to be stored at a certain temperature, interruptions in electricity may cause spoilage of the same. For an interruption of one minute the CDF is only Rs. 4.19/kw, but it goes up substantially as longer durations cause spoilage of perishables. If the interruption is as long as 8 hours, the CDF goes up to Rs. 328.50/kw. (Fig. 5.15).

TABLE 5.23
Cost Damage Function—Chemists

Interruption Duration	*CDF*
1 min.	4.19
20 min.	10.20
1 hr.	32.90
2 hr.	92.16
4 hr.	201.08
8 hr.	328.50

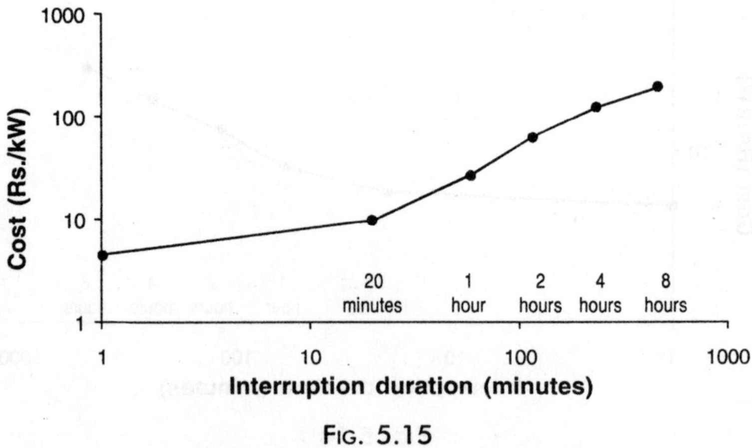

FIG. 5.15
Cost Damage Function—Chemists

Cost Damage Function—Cable Operators

The major cost, contributing to the damage function for this category is damage to equipment. Most of the organisations report using sensitive electronic equipment, which gets damaged with sudden interruptions. That is why even for an interruption of one minute the CDF is Rs. 26.37/kw. Another cost is associated with overtime costs when maintenance work has to be undertaken after an interruption has occurred, to resume the supply. For an 8 hr. interruption the CDF is Rs. 51.32/kw (Fig. 5.16).

TABLE 5.24
Cost Damage Function—Cable Operators

Interruption Duration	*CDF*
1 min.	26.37
20 min.	28.54
1 hr.	31.45
2 hr.	35.15
4 hr.	41.46
8 hr.	51.32

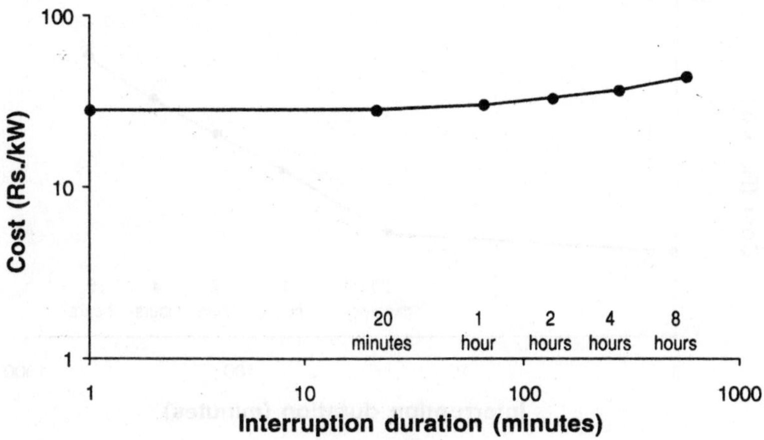

FIG. 5.16
Cost Damage Function—Cable Operators

Cost Damage Function—Beauty Parlours

The major cost contributing factors in this category are loss of sales and cost of running back-up. For shorter interruptions the CDF is not much, this being Rs. 2.79/kw for a one minute interruption and Rs. 3.75/kw for a 20 minute interruption. Loss of sales occur when the interruptions are of longer durations. For an 8 hour interruption the CDF is Rs. 75.16/kw. (Fig. 5.17)

TABLE 5.25
Cost Damage Function—Beauty Parlours

Interruption Duration	*CDF*
1 min.	2.79
20 min.	3.75
1 hr.	11.08
2 hr.	19.84
4 hr.	35.95
8 hr.	75.16

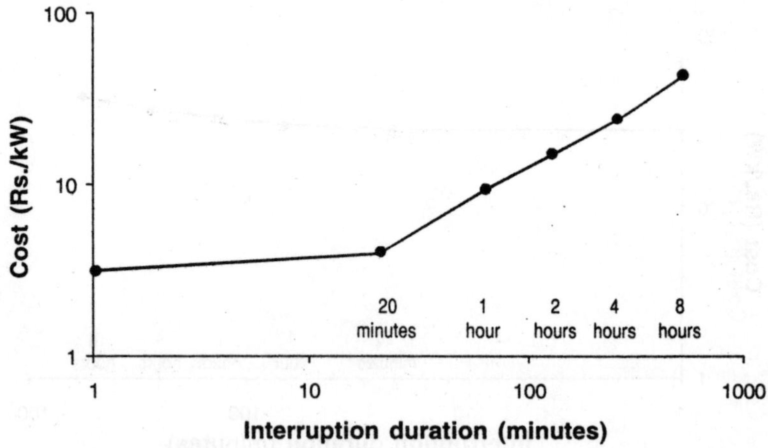

FIG. 5.17
Cost Damage Function—Beauty Parlours

Commercial Cost Damage Function

After calculating the cost damage function for all the customer classes, an overall cost damage function for the whole sector in the state has been worked out. The commercial cost damage function for Punjab is as shown by Table 5.26.

TABLE 5.26
Commercial Cost Damage Function

Interruption Duration	*CDF*
1 min.	16.38
20 min.	20.21
1 hour	36.12
2 hours	54.60
4 hours	84.34
8 hours	111.15

As the table indicates for an interruption duration of 1 minute the whole sector suffers a loss of Rs. 16.38/kw and Rs. 20.21/kw for an interruption of 20 minutes. This cost rises with increase in interruption duration, it being Rs. 111.15/kw for an interruption of 8 hours (Fig. 5.18).

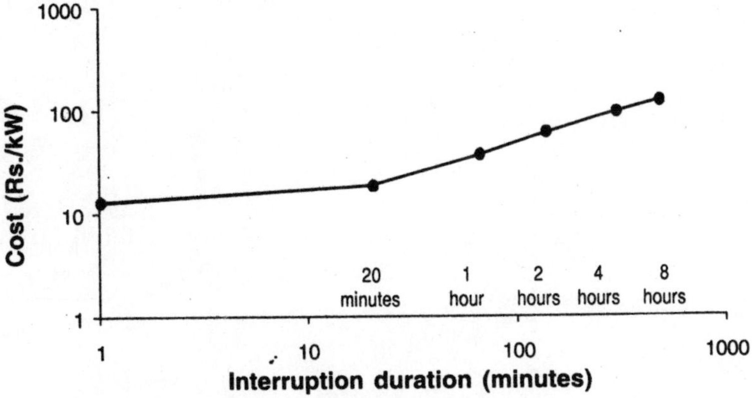

FIG. 5.18
Commercial Cost Damage Function

Note and Reference

1. R. Billinton, M. Pandey, "Reliability Worth Assessment of Electric Power Systems in Developing Countries", proceedings of PMAPS Conference, Vancouver, 1997.

6

Conclusions and Recommendations

CONCLUSIONS

1. In the commercial sector of Punjab, average number of electricity interruptions are of the order of 293 per year. Out of these, 22 interruptions last for four hours or more.
2. (i) Electrical equipment is of vital importance to all the organisations in this sector. The importance of various electrical equipments depends upon the type of commercial organisation.
   (ii) Air-conditioning, ventilation, refrigeration, indoor and outdoor lighting and some specialized equipment form the most vital equipment.
3. (i) 88% of the commercial organisations in Punjab make use of some back-up system or the other in their organisations, while the rest 12% have no back-up system.
   (ii) 20% commercial organisations have both engine generating systems and battery systems.
   (iii) 64.67% organisations have only generating sets in their organisations and 3.33% have only battery systems.

4. 38.67% of those using back-up systems, have systems which serve all their equipment. The rest 61.33% have systems which can run part of their equipment.
5. (i) 29% commercial organisations in Punjab foresee some health/safety risk associated with electricity interruptions, to their consumers, employees or public.
   (ii) All those organisations which foresee risk, would be able to reduce this risk if an advance warning about the impending interruption in provided.
6. (i) 58.67% of the commercial organisations in Punjab consider all months the same in terms of losses to their organisations, because of these interruptions. For the rest 41.33%, summer months, namely May, June, July and August are critical, June being the worst for most of commercial consumers (92.34%) for an interruption to occur.
   (ii) 66.33% commercial organisations consider all days the same, in respect of losses due to electricity interruptions. Another 20% perceive all week days to be the same and for nearly 70% of the rest Sundays are the worst. Thus, generally, weekdays are the worst for an electricity interruption with Saturday being the worst day for most of the commercial consumers (93.66%).
   (iii) For almost 35% commercial organisations, all times of the night and day cause the same amount of loss. For 21% all working hours are the worst, the working hours generally being 9 a.m. to 9 p.m. The rest have identified different time periods as the worst, the majority identifying 5 p.m. to 9 p.m. as the worst time. Hence for a majority of the organisations, 5 p.m. to 9 p.m. has been identified as the worst period for an interruption to occur (86.66%).
7. (i) For interruptions durations ranging from 20 minutes to 24 hours, the commercial organisations in Punjab, incur loss of sales ranging from Rs. 28.56/kw to Rs. 139.56/kw.
   (ii) Loss due to spoilage of perishables is upto Rs. 54.20/kw for an interruption of 24 hours.

(iii) The cost of running back up systems range from Rs. 1.08/kw to Rs. 42.29/kw for interruptions ranging from 20 minutes to 24 hours.

(iv) Total cost to the whole commercial sector ranges from Rs. 44.03/kw to Rs. 256.07/kw for interruption durations ranging from 20 minutes to 24 hours.

8. Seasonal variations affect the interrutpion costs. Costs have generally been found to be more in summers for majority of the organisations.
9. (i) Some commercial organisations can reduce these costs if an advance warning about an interruption is provided.

   (ii) A few organisations would be able to reduce their interruption costs if the information regarding the duration of the interruption is provided just at the start of the interruption.
10. The commercial cost damage function ranges from Rs. 16.38/kw for an interruption of one minute to Rs. 111.15/kw for an interruption of 8 hours.

RECOMMENDATIONS

1. It would be desirable if some steps are taken to reduce the number of interruptions. Some infrastructural/organisational changes must be made by the Punjab State Electricity Board to curtail the number of interruptions, and also to minimize the interruption duration.
2. An advance warning of the electricity interruptions must be provided in case of planned interruptions.
3. Some categories of commercial organisations face health/safety risk due to electricity interruptions. In case of unplanned outages, arrangements for alternate supply must be made to those identified classes of organisations.
4. Maintenance work and load shedding should be done keeping in view the worst periods identified by the study.

(i) Maximum maintenance work should be done in months prior to or after the summer months.

(ii) Any outage, if it is a planned one, should not occur during evening hours, which is the worst period for an electricity interruption, for the commercial sector.

5. Punjab State Electricity Board has a total commercial connected load of 9,02,633 kws. The commercial sector suffers a loss ranging from Rs. 147.9 lacs to Rs. 1003 lacs for an interruption duration lasting 20 minutes to eight affecting the whole of Punjab.
   If alternate parallel circuits are created to reduce the interruptions from the present 293/year, this loss incurred by the sector can be reduced. Cost incurred to create this redundancy in the distribution system by having alternate power supply arrangements, can easily be recovered by saving these losses.
6. CCDF can be converted to an Interrupted Energy Assessment Rate (IEAR) in order to incorporate the customer costs in planning further studies. The approach has already been used in many countries. These outage cost profiles can be used in the Board's systems planning and decision making.
7. If similar cost profiles for other sectors, i.e., industrial, agricultural and domestic are obtained, these can provide a composite cost damage function not only for Punjab but for other states in the country. This would help in undertaking cost benefit analysis to develop better power supply systems.
8. Assessment of reliability cost and reliability worth is an important element in the economic resource planning of any power generating organisation. Considerable work has been done in developed countries to optimize reliability levels of electric power systems on the basis of reliability cost and reliability worth. If reliability cost/worth approach is used by PSEB, it would provide an opportunity to planners to make decisions objectively and would therefore lead to an optimum expansion plan for the system.

7

Summary

INTRODUCTION

In modern society the pattern of social and working habits has made mankind wholly dependent on electricity. Moreover, the society as a whole expects the electric supply to be continuously available. Electricity has great significance for people across cultures and countries. Any interruption in the supply of electricity causes not only inconvenience but also certain tangible impacts.

The basic function of a power system is to supply customers, both large and small, with electrical energy as economical as possible and with an acceptable degree of reliability and quality. But this is not always possible due to random system failures-which sometimes are beyond reasonable control of power system managers. Many a times, local disturbances, adverse weather operational failure etc. cause the system to fail. All these failures add to the unreliability of electricity supply.

This unreliability of supply has different effects on different types of consumers. These consumers can be broadly classified into four categories—industrial, commercial, agricultural and residential. The impacts and costs of power failures to each sector are typically different from those in others.

Impact evaluation requires identification and quantification in monetary terms. Many direct impacts are relatively easy to identify and quantify, while others might be intangible (e.g. risk to life). The study of impacts of power interruptions is an important concept as it can be used in a cost-benefit analysis for developing/improving the power systems. This economic analysis is fundamental to establish a balance between expenditure of obtaining a certain level of reliability and the worth of having that level.

The present research project, is aimed at studying the impact of electricity interruptions on the commercial sector in Punjab.

Commercial sector is a vital part of the economy of any state/country. Its well being is important for the well being of the entire society. Punjab is one of the most progressive states in India, having been a pioneer in green revolution. The only organisation which generates and supplies electricity in Punjab is the Punjab State Electricity Board, an autonomous body.

The Board supplies electricity to all categories of consumers in the state, having distinct classification of electricity connections and tariff structures across consumer classes. The commercial sector, (designated as non-residential supply connections) is amongst the highest tariff paying classes in Punjab.

It is assumed that electricity interruptions would have certain impact on this sector, too, as in case of other sectors. Through this research an endeavour is being made to identify and quantify that impact in both tangible and intangible terms.

For the purpose of the present research;
Impact means—effect/consequence to consumer.
Interruptions mean—any outage, planned or unplanned.

Commercial sector means—Organisations with an electricity connection categorized as NRS (non-residential supply) by PSEB.

Objectives of the Study

The following, briefly are the aims and objectives of the study :

1. *To find out the perception of commercial consumers regarding reliability of supply*

The first objective of the research would be to get an estimate of number of power interruptions from the respondents. This would be done to get an idea of the frequency and duration of electricity interruptions.

2. *To find out the importance of electric equipment to commercial organisations*

To study the impact of electricity interruptions on consumers, it is important to establish their dependence on electricity. Thus, the objective would be to find out the criticality of electric equipments/gadgets to commercial consumers. This would establish the level of their dependence on electricity and hence the cost of unreliability of supply.

3. *To find out the extent of use of back-up systems by commercial organisations*

In case a consumer is critically dependent on electricity and suffers a great loss due to unreliability of supply, then some alternate arrangements would definitely be made to minimize the impact. As a sequence to the first and the second objectives of the study, i.e., to establish the level of unreliability and the dependence of consumers on electricity, the next objective would be to find out the level and extent of alternate arrangements. Generating sets are the most commonly used alternative as back-up. The aim here would be to identify the extent of use of generating systems by the commercial sector, and also the capacity of the generating sets being employed. The capability of the generating sets to run various equipments being used by the consumers would also be established.

4. *To investigate if any health/safety risk is posed by electricity interruptions*

One of the intagible impacts of electricity interruptions can be health/safety risk to the employees and customers of commercial organisations or to the general public. The objective would be to establish whether an interruption in electricity causes any such risk.

In cases where such a risk is established, it would be further investigated if an advance warning of the impending interruption can reduce the risk.

5. *To identify the worst time for an interruption to occur*

Since organisations vary in their size, location, area of operations etc., the time period when an interruption has the worst impact, may also be variable. The aim would be to identify the worst time for an interruption to occur, i.e. when the organisation would suffer the most losses due to an electricity interruption.

This would include establishing the worst month of the year, the worst day of the week and the worst time during the day for commercial organisations.

6. *To identify the monetary impact of electricity interruptions*

The major impact of electricity interruptions is the monetary loss suffered by organisations as a consequence of interruptions. The aim would be to identify and estimate that monetary loss.

The estimate would be for different durations of electricity interruptions ranging from two seconds to twenty-four hours.

7. *To study the effect of seasonal variations on impact of electricity interruptions*

The aim would be to identify any variation in the losses suffered by commercial consumers due to variation in seasons. Since Punjab has extreme seasonal variations, fluctuating between severe summers and severe winters, it would be identified if these changes bring about any change in the cost profiles of commercial consumers due to interruptions.

8. *To study the effect of advance warning on losses due to electricity interruptions*

The objective would be to find out whether providing an advance warning about an impending interruption to the commercial consumers would help consumers in reducing their losses. It would also be established that in case the consumers are able to reduce their losses, what would be the extent of that reduction in loss and also as to how much advance warning would be required for the same.

Another objective here would be to find out whether providing information regarding the duration of interruption to the consumers just after the interruption has started would help in cost reduction. Again, information would be collected about the interruption duration, which if intimated at the commencement of the interruption, would reduce the loss.

9. *To Establish Commercial Cost Damage Function (CCDF)*

The most direct and usable form of customer perceived interruption costs is customer damage function. The CDF portrays the costs associated with supply interruption as a function of outage duration.

The objective would be to find out the CCDF, and also CDF's of classes within the sector, since CDF can also be calculated for a particular class of customers.

SURVEY OF RELATED LITERATURE

In their research paper published in IEEE Transaction on Power Apparatus and Systems in Oct. 1983, G. Wacker, E. Wojczynski and R. Billinton have discussed the results of a residential survey. It was carried out to study the costs incurred by residential consumers due to electricity interruptions. The study used a mail survey to determine the residential consumers' assessment of the cost of electric service interruptions. The survey obtained user's cost evaluation using three approaches. Two of these approaches asked respondents to indicate rate changes that would be commensurate with specified changes in reliability, while the third approach was an indirect worth evaluation based on the preparatory actions a consumer predicted he would take, given a specific level of unreliability. User's evaluations are reported for a number of frequency and duration failure scenarios.

The major contribution of this work is the compilation of residential cost of interruption duration as a function of both user and interruption characteristics. It has also contributed to a great extent in developing interruption costing methodology.

The same authors, in Februry 1984 issue of IEEE Transactions, present 'Interruption Cost Methodology and Results—A Canadian Commercial and Small Industry Survey'. This paper presents the results of an investigation of the direct short term impacts and costs incurred by commercial and small industrial customers resulting from local random supply interruptions. The study used a mail survey to determine the customer assessment of electric service interruptions. The variation in interruption cost with the variation in characteristics of interruptions such as duration of the interruption, time of the day, day of the week and month of the year, advance warning and reliability history and with the characteristics of the users was also investigated.

The paper has presented a summary of research conducted to determine estimates of the perceived costs or losses to commercial and small industrial consumers.

In July, 1990 issue of IEEE Transactions—Power Engineering Review, Daniel J. World, published his paper 'Power Quality—Two different Perspectives'. This paper focuses on the problem of short duration transients or momentary disturbances and their effect on sensitive customer equipment like computers. One perspective, from the utility side of the meter, covers any power system disturbances which originate on that side of the meter and which adversely affect customers' equipment. The other perspective is from the customer side of the meter. In that regard, the causes of the disturbances are usually restricted to the customers' premises.

The author points out that many equipments like minicomputers, electronic cash registers and data terminals etc. are affected because of their sensitive circuitry and cause memory loss, system malfunction and even component failure.

The paper on 'Interruption costs, customer satisfaction and expectations for service reliability' by Michael J. Sullivan and B. Noland Suddeth (U.S.A.) summarizes results of a comprehensive study of the economic value of electric service carried out by Duke Power Company, Charlotte, U.S.A. in cooperation with Electric Power Research Institute. In the study, customer interruption costs were estimated for generation, transmission and distribution outages of differing lengths occuring under varying circumstances. Interruption costs for momentary outages and voltage disturbances are also reported.

The basic methodology used in this study involves directly asking random samples of customers in different market segments (i.e. residential, commercial and industrial) to estimate their economic losses as a result of power reliability and quality problems.

Sullivan and Suddeth find that interruption costs experienced by commercial and industrial customers are much higher than residential customers. Further they conclude that interruption costs vary from customer to customer depending on a number of factors, like the customers business type and size. Using these production factors, they developed multiple regression models for predicting customer interruption costs.

Regarding customer expectations for service reliability, the authors point out that customers differ dramatically in their expectations. Commercial and industrial customers expect nearly perfect reliability. Residential customers have significantly lower expectations for service reliability.

This study shows that customer interruption costs vary systematically and predictably as a function of customer type and size and within commercial and industrial customers by the process, equipment and products being made and sold. It also documents the effect of advance warning on interruption costs arising from generation outages and suggests that electric emergency planning may be a highly cost effective alternative to investment in new generation.

X-Liang and L. Goel of Nanyang Technological University Singapore, in 1996, published a research paper, 'Distribution System reliability evaluation using the Monte Carlo simulation method.'

The authors point out that reliability evaluation of individual customer service continuity levels is an important aspect which can significantly affect the design and operating characteristics of power systems in general, and distribution circuits in particular. Analysis of customer failure statistics indicates that distribution systems make a significant individual contribution to overall customer supply unavailability. Statistics such as these reinforce the need to conduct quantitative adequacy assessments not only for average indices but also for their distributions in the distribution functional zone.

This paper describes a Monte Carlo approach to a comprehensive distribution system reliability evaluation. The paper presents the results of load point reliability indices and system performance indices for six cases using the Monte Carlo method for a distribution test system.

In particular, the probability distributions of system performance indices such as SAIFI, SAIDI and CAIDI for the six cases are presented. The probabilities of these indices occurring during different intervals are analyzed and the effects of different operation philosophies on the system performance indices are illustrated from the probability distribution point of view. The results obtained by the simulation method are compared with those obtained using an analytical approach. This paper illustrates that Monte Carlo simulation is a useful and powerful tool for comprehensive distribution system reliability assessment. The study results show that the simulation method gives acceptable reliability indices and can also be used to provide information on the probability distributions associated with the predicted indices.

R. Billinton and M. Pandey, in their paper, 'Customer Interruption Cost Assessment : An International Perspective', published in 1999, present some interruption cost profiles obtained from quite diverse jurisdictions of the world.

The authors say that assessment of reliability has become a well established utility practice. Customer interruption costs are considered to be a surrogate for the actual worth of service reliability and widely used in its assessment.

The most common index used to link a probabilistic approach with customer interruption costs is the expected energy not supplied (EENS). This can be predicted for generation, composite generation and transmission and distribution systems using established techniques. Customer interruption cost can be expressed in terms of cost per unit of unserved energy designated as the interrupted energy assessment rate (IEAR) or the value of lost load (VOLL). The product of EENS and IEAR (VOLL) provides an explicit monetary value associated with the system condition under study and can be used in a wide range of utility decision-making process.

The authors point out that though a variety of methods have been utilized to estimate the impacts of power outages on electricity consumers, the customer survey approach seems to be the most popular and practical. The most direct and usable form of customer percieved interruption costs is the customer damage functions (CDF). The CDF portrays the costs associated with supply interruptions as a function of outage duration. The costs are usually normalized by dividing the actual customer interruption costs by either the annual energy consumed or the peak load.

The paper presents the customer interruption cost profiles obtained using surveys in many countries. The results show that although the interruption costs differ in magnitude, these are large in developed countries as compared to developing countries, but the cost profiles obtained are very similar.

RESEARCH METHODOLOGY

Sample Design

The first step in the sampling process is defining the population. For the present study all the electricity consumers in the state of Punjab, having a commercial connection constitute the population. Information regarding both, the number of commercial connections and the total connected load (commercial) was obtained for all the cities in the state. The four cities, Jalandhar, Ludhiana, Amritsar and Patiala had the highest commercial loads, so these cities were selected for the study. These cities not only have the highest number of commercial load, but also the highest number of commercial connections in the whole state and constitute 40% of the population. The following table shows the number of commercial connections and the connected load (comm.) in these four cities.

In order to draw a sample, stratified random sampling technique has been used. The reason for taking a stratified sample is to have a more efficient sample than could be taken on the basis of simple random sampling. Random sampling error

TABLE 7.1
Total Commercial Connections in Punjab (Population of the study)

City	*No. of Commercial Connections*	*Comm. Connected load (KW)*
Patiala	39,368	77,917
Amritsar	46,928	84,150
Ludhiana	64,199	1,44,245
Jalandhar	58,712	1,16,473
Total Punjab	5,33,717	9,02,633

will be reduced, because the groups are internally homogenous, comparatively different amongst themselves. A smaller standard error may be the result of this stratified sample because the groups will be adequately represented when strata are combined. Moreover, occasionally a simple random sample yields a disproportionate number of one group or another and the representatives of the sample could be improved by this sampling method.

Size and structure is not the only characteristic but the area of operation of commercial organisations play a major role for this study. The difference in the field of operations, makes some organisations more susceptible to electricity interruptions than others. Hence to include all types of business in the sample, the population has been divided into seventeen stratas. (See Table 7.2 on next page).

A total of 300 respondents have been taken, covering all the stratas/classes, taking 75 respondents from each city.

Questionnaire Design

Research tool used for the present study is a questionnaire. After a careful analysis of all the objectives of the study exercises were undertaken to construct and word appropriate questions. Many tools used by earlier researchers in the related areas were studied.

'Power Systems Research Group' of Department of Electrical Engineering, University of Saskatchewan, Canada has

TABLE 7.2
Strata-wise Number of Respondents

City	*No. of Commercial Connections*	*Comm. Connected load (KW)*
Nursing Homes	19	6.33
Cinemas	20	6.67
Restaurants	30	10.00
Shops/Showrooms	27	9.00
Petrol Pumps	23	7.67
Hotels	15	5.00
Diagnostic Labs.	15	5.00
Motor Garages	14	4.67
Colour Labs.	14	4.67
Dairies	19	6.33
Marriage Palaces	15	5.00
Dry Cleaners	15	5.00
Computer Centres	17	5.67
Printing Presses	14	4.67
Chemists	16	5.33
Cable Operators	14	4.67
Beauty Parlours	13	4.33
Total	300	100

been continuously conducting research on electricity interruptions in Canada. A questionnaire developed by the group for their study, 'Assessment of Reliability worth in Electric Power Systems in Canada' in 1993 has been adopted for the present study. This questionnaire was modified and restructured to suit Indian population/conditions.

This questionnaire was then tested for its validity and reliability.

Validity refers to the concept, that the tool actually measures what it intends to measure.

The present questionnaire's validity was measured by using the method of content validity. Face or content validity refers to the subjective agreement among professionals that a scale logically appears to be accurately reflecting what it propose to measure. And that the content of the scale appears to be adequate. When it appears evident to experts that the measure provides adequate coverage of the concept, the measure has face validity. So, the questionnairre was shown to various academicians and also other experts for checking its content validity. A number of modifications were made on the basis of the suggestions received from these experts.

Reliability refers to the concept that when the outcome of the measuring process is reproducible, the measuring instrument is reliable. Reliability applies to a measure when similar results are obtained over time and across situations. The two dimensions underlying the concept of reliability are : repeatability and consistency. To measure the reliability of the present scale, the test-retest method was used. It involves administering the same scale to the same respondents at two separate points in time to test for stability. The two sets of scores obtained are then checked for their correlation. In the present case a correlation score of .90 was obtained when the scale was administered to the same respondents at two different points of time.

Hence questionnaire was found to be valid and reliable to be used for the present research project.

Collection of Data

Survey—also known formally as a sample survey method of data gathering has been adopted. Survey have an advantage over other methods, as these provide a quick, inexpensive, efficient and accurate means of gathering information about the population. A review of the literature reveals that investigations into the cost of power interruptions have been conducted using a variety of approaches. The approach which is considered to yield relatively definitive results in the assessment of direct costs, particularly for such customers as commercial and industrial users, is the customer survey approach. Hence, it was decided to use survey to conduct the study. The basis of this approach is that customers are in the best position to understand power interruption consequences in relation to their particular needs.

Every respondent in the present sample was contacted personally to get the questionnaires filled. Personal contact with the respondents facilitates the collection of accurate data, as any misunderstanding on the part of the respondent is cleared by the researcher on the spot.

Data for the present project was collected through personal contact during the summer months of the year 1999. The respondents were contacted during the day, at their place of work. Subsequent visits were made if questionnaire could not be filled during one visit due to busy schedule of the respondents. In case of all the 300 respondents either owner/ partner or manager was asked to answer the questions to ensure the accuracy and reliability of the information.

Data Analysis

To analyse the data, in order to get meaningful information (results) regarding the impact of electricity interruptions on commercial consumers, mainly percentages, averages have been calculated. A particular effect of an interruption has been studied as to how much percentage of respondents have been affected.

Direct costs estimated by respondents have been calculated in a similar fashion. Costs indicated by respondents for specific durations of interruptions have been averaged for all the categories.

Cost Normalization

The primary purpose of conducting this study is to establish monetary losses associated with various levels of unreliability. These costs must be in a form that is most usable to the organisation supplying electricity. A simple average or mean value of interruption costs can create a distorted understaning of the actual costs associated with an interruption. Therefore the costs have been normalized by using the connected loads of the consumers. The procedure used to normalize costs is an 'aggregated average' approach. The interruption costs associated with the respondents are totalled and the respondents' connected loads are also totalled. The ratio of the sum of the costs and the sum of the connected loads is designated as an 'aggregated average'. This procedure produces a different result than if the individual users' costs were normalized and then averaged. the aggregated averages approach is used to off-set the impact of small numbers of large or small consumers, or small numbers of respondents who report large or small costs.

Commercial Cost Damage Function

The most direct and usable form of customer perceived interruption costs is the customer damage function.[8] It portrays the costs associated with outages as a function of outage duration. The CDF can be determined for a particular customer category or for an overall sector. Commercial Cost Damage Function has been calculated for the sector and also for classes within the sector by weighting the aggregated customer costs by the percentage of their load mixes and then summing these costs.

IMPACT OF ELECTRICITY INTERRUPTIONS

(i) Number of Interruptions

Table 7.3 shows the number of failures reported by respondents during the last one year. As is shown in the table, the average number of interruptions reported by commercial sector respodents are 292.75. The average number of interruption lasting 4 hours or more was 21.65.

TABLE 7.3
Number of Failures in Last One Year

City	*No. of Commercial Connections*	*Comm. Connected load (KW)*
0	0	0
1-100	95	31.67
101-200	84	28.00
201-300	45	15.00
301-400	20	6.67
401-500	16	5.33
501-600	12	4.00
More than 600	28	9.33
Average no. of failures	292.75	

(ii) Importance of Electric Equipment

Respondents were asked to check, which electric equipment was important to their organisations for their day-to-day operations. A list of electric equipment were provided in the questionnaire along with a choice of 'any other', so that if any equipment, not listed, was important to the respondent, the same could be mentioned there.

Equipment like ventilation, airconditioning, refrigeration, indoor and outdoor lighting were reported to be most important by a majority of the respondents. Some specialised equipment used by nursing homes, cinemas, cable operators etc. were also reported to be of utmost importance to the organisations. The responses show that the total operation of some businesses come to a standstill on account of electricity interruptions.

(iii) Use of Back-up Systems

Since all respondents have indicated unreliability of supply and great dependence on some electric equipment, they were further asked if they took any anticipatory actions in case of electricity interruptions.

Only 12% respondents had no back-up system, the rest 88% has either engine generators or battery systems to carry out the working of the organisations in the event of power failure.

As shown in the table, 20% respondents have both engine and battery systems, 3.33% have only battery systems and 64.67% have only engine generator.

(iv) Health/Safety Hazard

On being asked whether electricity interruptions posed any health/safety risk to the employees/customers of these organisations, 71% respondents have reported that interruptions pose no health/safety risk. The rest 29% perceive some risk to the employees/public in event of electricity interruptions.

Categories of respondents perceiving some risk to their employees are those dealing in medicine—namely, nursing homes, diagnostic labs and chemists. A few hotels, cinemas and shops/showrooms also foresee some risk, mainly due to stampedes etc.

TABLE 7.4
Back-up System

Category	*Battery System*		*Engine Generators*		*Both*		*No Back-up*	
	N	%	*N*	%	*N*	%	*N*	%
Nursing Homes	1	5.26	14	73.68	4	21.05	—	—
Cinemas	—	—	18	90.00	2	10.00	—	—
Restaurants	1	3.33	25	83.33	2	6.67	2	6.67
Shops/Showrooms	—	—	20	74.07	4	14.81	3	11.11
Petrol Pumps	—	—	16	69.57	7	30.43	—	—
Hotels	—	—	12	80.00	3	20.00	—	—
Diag. Labs.	—	—	10	66.67	4	26.67	1	6.67
Motor Garages	—	—	10	71.43	4	28.57	—	—
Colour Labs.	1	7.14	8	57.14	2	14.29	3	21.43
Dairies	1	5.26	5	26.32	—	—	13	68.42
Marriage Palaces	—	—	12	80.00	3	20.00	—	—
Dry Cleaners	—	—	10	66.67	1	6.67	4	26.67
Computer Centres	2	11.76	6	35.29	9	52.94	—	—
Printing Presses	—	—	4	28.57	5	35.71	5	35.71
Chemists	—	—	12	75.00	—	—	4	25.00
Cable Operators	2	14.29	3	21.43	9	64.29	—	—
Beauty Parlours	2	15.38	9	69.23	1	7.69	1	7.69
Total	10	3.33	194	64.67	60	20.0	36	12.0

(v) Worst Time for an Interruption

The respondent were asked to identify the worst month of the year, day of the week and time of the day for an interruption to occur.

(a) Worst Month

Of the total 300 respondents, 176, i.e. 58.67% said that for their business, all months are the same. The rest 41.33% respondents have selected different months, which account for more losses as compared to other months of the year.

Out of these 41.33% respondents, 81.4% have selected June as the worst month, followed by May by 75% and July by 67.7% respondents.

(b) Worst Day

Here, 66.33% respondents reported that for them all days are the same. Another 20% respondents have opted for 'all weekdays the same, i.e., excluding Sundays. 13.67% respondents have selected particular days and out of these, 70.70% respondents have chosen Sunday as the worst day, followed by Saturday by 53.60% respondents.

(c) Worst Time of the Day

Worst time of the day was also selected using the same methodology. 34.6% respondents have reported that for them, all times of the day and night are the same. 23.4% respondents have indicated all working hours the same, the average working hours being 9 a.m. to 9 p.m. 42% respondents have chosen particular time durations during the day, which they feel are worse than other times. Out of these 42%, 74.6% respondents have chosen 5 p.m. to 9 p.m. as the worst time, followed by 4 to 5 p.m. by 58.7% respondents.

(v) Direct Cost Evaluation

After the respondents identified the worst month, worst day and the worst time of the day for an interruption to occur, they were asked to assume that an interruption occured at the wors time chosen by them. They had to estimate losses to their organisation, due to the failure, assuming that the back-up system was working during the period of interruption. The factors contributing to loss were taken as :

(a) wages paid to idle workers.
(b) loss of sales.
(c) overtime costs.
(d) damage to equipment.
(e) spoilage of perishables.
(f) cost of running back-up.
(g) cost of any special procedures.

The time periods for which respondents were asked to estimate their losses are :

(a) 2 seconds.
(b) 1 minute.
(c) 20 minutes.
(d) 1 hour.
(e) 2 hours.
(f) 4 hours.
(g) 8 hours.
(h) 24 hours.

The total costs obtained are as shown in the following table.

(vii) Variation in Interruption Costs

The cost to commercial sector remains maximum in the summer months, i.e. from April to September. In rest of the months there is some reduction in the costs though a total of 197 respondents have said that for them all months are the same.

During the week, interruptions during weekdays came the most costs, with a reduction of cost on Sundays.

For various time durations, evenings are the worst, with most of the consumers saying that during morning hours, the cost is lesser as compared to evening.

(viii) Effect of Advance Warning

After the respondents identified the losses to their business, they were asked whether an advance warning would

Table 7.5
Total Costs

Categories	2 Second		1 Minute		20 Minutes		1 Hour	
	%	Avg. loss (Rs.)	%	Avg. loss (Rs.)	%	Avg. loss (Rs.)	%	Avg. loss (Rs.)
(1)	(2)	(3)	(4)	(5)	(6)	(7)	(8)	(9)
Nursing Homes	78.95	6112.11	78.95	6112.11	94.74	6157.75	100.00	8333.95
Cinemas	80.00	1002.50	80.00	1002.50	100.00	1056.75	100.00	3684.25
Restaurants	6.67	83.33	13.33	83.87	96.67	129.37	100.00	829.50
Shops/Shorooms	37.04	196.30	37.04	196.30	92.59	1430.78	100.00	3792.22
Petrol Pumps	30.43	82.61	30.43	82.61	100.00	127.87	100.00	198.26
Hotels	33.33	573.33	33.33	573.33	100.00	635.20	100.00	784.67
Diag. Labs.	66.67	586.67	66.67	586.67	93.33	602.67	100.00	890.13
Motor Garages	28.57	178.57	28.57	178.57	100.00	217.07	100.00	461.79
Colour Labs.	85.71	203.57	85.71	203.57	92.86	253.43	92.86	330.71
Dairies	10.53	5.26	10.53	5.26	42.11	23.42	84.21	167.37
Marriage Palaces	40.00	226.67	40.00	236.67	100.00	285.53	100.00	396.67
Dry Cleaners	6.67	6.67	6.67	6.67	73.33	23.93	93.33	100.33
Computer Centres	30.59	305.88	30.59	305.88	100.00	338.82	100.00	413.82
Printing Presses	42.86	132.14	42.86	132.14	78.57	155.79	78.57	226.43
Chemists	31.25	37.50	37.50	43.75	87.50	106.44	93.75	343.44
Cable Operators	85.71	685.71	85.71	685.71	92.86	742.07	92.86	817.64
Beauty Parlours	23.08	46.15	23.08	46.15	84.62	62.08	100.00	183.23
All Respondents	42.67	633.60	43.67	633.99	90.33	782.11	96.67	1463.82

Table 7.5 (Contd.)

Categories	2 hours		4 hours		8 hours		24 hours	
	%	Avg. loss (Rs.)	%	Avg. loss (Rs.)	%	Avg. loss (Rs.)	%	Avg. loss (Rs.)
(1)	(10)	(11)	(12)	(13)	(14)	(15)	(16)	(17)
Nursing Homes	100.00	8987.37	100.00	11328.42	100.00	13515.79	100.00	15455.79
Cinemas	100.00	4066.00	100.00	7279.50	100.00	7997.50	100.00	8539.25
Restaurants	100.00	2345.30	100.00	4059.00	100.00	5734.67	100.00	6792.67
Shops/Shorooms	100.00	6956.11	100.00	10535.56	100.00	14522.96	100.00	15593.35
Petrol Pumps	100.00	325.22	100.00	547.17	100.00	972.61	100.00	1832.17
Hotels	100.00	1156.00	100.00	2985.33	100.00	3044.00	100.00	5439.33
Diag. Labs.	100.00	1063.67	100.00	1624.40	100.00	2261.33	100.00	2867.33
Motor Garages	100.00	184.29	100.00	1197.14	100.00	1897.86	100.00	2311.07
Colour Labs.	100.00	475.71	100.00	626.43	100.00	913.57	100.00	1085.71
Dairies	100.00	436.05	100.00	558.42	100.00	780.00	100.00	1038.95
Marriage Palaces	100.00	566.67	100.00	986.67	100.00	1586.67	100.00	3370.67
Dry Cleaners	100.00	199.00	100.00	353.00	100.00	540.67	100.00	736.00
Computer Centres	100.00	527.65	100.00	749.41	100.00	1134.12	100.00	1472.06
Printing Presses	100.00	415.00	100.00	707.14	100.00	1167.14	100.00	1970.00
Chemists	100.00	961.88	100.00	2098.75	100.00	3428.75	100.00	4180.94
Cable Operators	92.86	913.93	100.00	1077.86	100.00	1334.29	100.00	2344.29
Beauty Parlours	100.00	328.08	100.00	594.62	100.00	1243.08	100.00	1694.23
All Respondents	99.67	2081.57	100.00	3204.89	100.00	4241.57	100.00	5124.08

reduce their interruption costs. The following results were obtained.

TABLE 7.6
Reduction in Interruption Costs (Advance Warning)

Categories	*% able to Reduce Cost*
Nursing Homes	36.84
Cinemas	15.00
Restaurants	30.00
Shops and showrooms	14.81
Petrol Pumps	8.70
Hotels	13.33
Diag. Labs	46.67
Motor Garages	7.14
Colour Labs	71.43
Dairies	42.11
Marriage Palaces	13.33
Dry Cleaners	40.00
Computer Centres	23.53
Printing Presses	42.86
Chemists	37.50
Cable Operators	14.29
Beauty Parlours	—
All Respondents	26.33

As shown in the table, 26.33% respondents would be able to reduce their costs, if they have a knowledge of the impending interruption.

Similarly, information was obtained regarding reduction in costs, when information regarding the length of the interruption is provided once the interruption has started. Here only 5% respondents have indicated a reduction in costs.

COST NORMALISATION AND CCDF

After getting direct cost estimates from all the respondents for durations ranging from 2 seconds to 24 hours, their connected loads were obtained. The costs provided by each respondent were normalized with respect to their connected loads to get cost in Rs./kw. The average cost due to interruption to each group of respondents have yielded cost/kw. for the period of interruption.

Such normalized costs have been obtained for all categories of respondents. The costs have also been obtained for various factors (e.g. wages paid to idle workers, spoilage of perishables etc.).

All the costs which have been dealt individually, make up total cost figures which are the total normalized costs of electricity interruption. These are shown in the following Table 7.7.

Commercial Cost Damage Function

The most direct and usable form of customer perceived interruption costs is the customer damage functions (CDF) (R. Billinton, M. Pandey). The CDF indicates the costs associated with power interruption as a function of outage duration. The CDF can be determined for a particular customer class within a sector or for an overall customer sector.

For the purpose of the present study, the CDF is calculated for all the customer classes included in the study and also for the whole sector, which would be called commercial cost damage function (CCDF).

CCDF for Punjab is shown by the Table 7.8 on next page.

This table indicates that for an interruption duration of 1 minute, the commercial sector consumers suffer a loss of Rs. 16.38/kw and for an interruption of 8 hours the cost would be Rs. 111.15/kw.

CONCLUSIONS AND RECOMMENDATIONS

Conclusions

1. On an average, 293 electricity interruptions occur per year in the commercial sector of Punjab.
2. Electrical equipment is of great importance to

TABLE 7.7
Total Cost

Categories	*2 sec.*	*1 min.*	*20 min.*	*1 hr.*	*2 hrs.*	*4 hr.*	*8 hr.*	*24 hr.*
Nursing Homes	76.18	76.18	77.10	124.27	140.52	182.40	229.55	280.11
Cinemas	9.30	9.30	9.82	35.75	41.48	92.46	109.20	114.77
Restaurants	5.86	5.92	286.17	335.29	459.99	653.37	879.19	942.58
Shops/Showrooms	12.96	12.96	19.12	103.59	223.50	335.88	447.36	483.44
Petrol Pumps	7.72	7.72	10.28	13.31	20.42	32.76	54.78	100.96
Hotels	12.15	12.15	13.03	15.34	23.50	50.06	51.14	89.43
Diagnostic Labs	21.97	21.97	22.96	66.45	78.51	126.19	180.59	244.51
Motor Garages	6.43	6.43	8.07	26.32	44.74	70.63	112.63	126.59
Colour Labs	19.85	19.85	22.66	31.36	48.28	60.44	81.89	95.22
Dairies	1.10	1.10	4.71	29.32	75.98	97.22	137.11	172.44
Marriage Palaces	3.22	3.22	4.03	5.56	7.91	12.59	21.96	45.53
Dry Cleaners	0.39	0.39	1.74	7.46	14.74	25.56	37.81	49.60
Computer Centres	13.49	13.49	15.09	18.37	23.45	33.41	51.37	64.20
Printing Presses	3.27	3.27	4.39	8.04	18.78	31.58	55.19	94.40
Chemists	3.85	4.63	10.20	34.86	97.24	213.59	357.40	429.26
Cable Operators	33.69	33.69	38.11	43.38	49.52	58.56	69.51	110.95
Beauty Parlours	2.22	2.22	3.69	11.42	21.40	38.82	72.34	146.08
All Respondents	13.76	13.81	44.03	69.88	105.36	157.96	217.46	256.07

TABLE 7.8
Commercial Cost Damage Function

Interruption Duration	*Cost/kw (Rs.)*
1 min.	16.38
20 min.	20.21
1 hour	36.12
2 hours	54.60
4 hours	84.34
8 hours	111.15

commercial organisations. The importance of electrical equipments depends upon the type of organisation.

3. 88% of commercial organisations in Punjab make use of back-up systems.
4. Almost 29% commercial organisations foresee some health/safety risk to their customers, employees or public due to electric interruptions. This risk can be reduced if an advance warning is provided.
5. Seasonal variations affect the interruption costs.
6. Advance warning about an interruption helps in reducing these costs.
7. The commercial cost damage function ranges from Rs. 16.38/kw for an interruption of one minute to Rs. 111.15/kw for an interruption of 8 hours.

Recommendations

1. Some infrastructural/organisational changes must be made by PSEB to reduce the number of interruptions and to minimise the interruption duration.
2. An advance warning of the planned electricity outages must be provided whenever possible.
3. Alternate supply arrangements must be made for the organisations facing health/safety risk.
4. Maintenance work and load shedding should be avoided during the worst periods identified in the study.
5. Commercial cost damage function must be utilized to plan cost-benefit studies for having alternate power supply arrangements.

Bibliography

A Report Proposal by the Reliability Test System Task Force of the Application of Probability Methods Sub-committee, The IEEE Reliability Test System—1996, IEEE Transactions on Power Systems, Vol. 14, No. 3, August 1996.

Allan, R., *Power System Reliability Assessment—A Conceptual and Historical Review,* Reliability Engineering and System Safety 46, 1994.

Barakat, E.H. and Al-Rashed, S.A., *Analysis of the Impact of Dynamic Events on Future Power Requirements,* Electric Power Systems Research 37, 1996.

Billinton, R. and Allan, R.N., *Guest Editorial,* Reliability Engineering and System Safety 46, 1994.

Billinton, R. and Allan, R.N., *Reliability Assessment of Large Electric Power Systems,* Kluwer Academic Publishers, Boston, 1988.

Billinton, R. and Allan, R.N., *Reliability Evaluation of Engineering Systems, Concepts and Techniques,* Longman, London (England)/Plenum Publishers, New York, 1983.

Billinton, R. and Allan, R.N., *Reliability Evaluation of Engineering Systems : Concepts and Techniques,* Plenum Publishing, New York, USA, 1992.

Billinton, R. and Allan, R.N., *Reliability Evaluation of Power Systems,* Longman, London (England)/Plenum Publishers, New York, 1984.

Billinton, R. and Allan, R.N., *Reliability Evaluation of Power Systems,* Pitman, New York, 1984.

Billinton, R. and Pandey, M., *Customer Interruption Cost Assessment : An International Perspective,* IEEE Power Engineering Review, January 1999.

Billinton, R. and Peng Wang, *Teaching Distribution System Reliability Evaluation Using Monte Carlo Simultation,* IEEE Transactions on Power Systems, Vol. 14, No. 2, May 1999.

Billinton, R., Allan, R.N. and Salvaderi, L., *Applied Reliability Assessment in Electric Power Systems,* IEEE Press, 1991.

Billinton, R., and Pandey, M., *Reliability Worth Assessment of Electric Power Systems in Developing Countries,* Proceedings of PMAPS Conference, Vancouver, 1997.

Billinton, R., *Evaluation of Reliability Worth in an Electric Power System,* Reliability Engineering and System Safety 46, 1994.

Billinton, R., Wacker, G. and Wojczynski, E., *Customer Damage Resulting from Electric Service Interruptions,* Canadian Electrical Association, R and D Project 907 U 131 Report, 1982.

Billinton, R., Wacker, G. and Wojczynski, E., *Customer Damage Resulting from Electric Service Interruptions,* (Vol. I) CEA R&D Project, 1982.

Billinton, R., Wacker, G. and Wojczynski. E., *Power System Reliability Indices and their Application in the Assessment of Reliability Worth,* CEA Transactions, Vol. 21, 1982.

Ching-Tzong Su, and Guor-Rurng Lii, *Reliability Planning for Composite Electric Power Systems,* Electric Power Systems Research 51, 1999.

Cho, M.Y., Chen, C.S., and Lin, W.M., *Design of a Distribution Database with Customer Daily Load Patterns,* Electric Power Systems Research 1992.

Crozier, J.T. and Wisdom, W.N., *A Power Quality and Reliability Index based on Customer Interruption Costs,* IEEE Power Engineering Review, April 1999.

Daniel J. Ward, *Power Quality—Two Different Perspectives,* IEEE Power Engineering Review, July 1990.

Debnath, K. and Goel, L., *Power System Planning—A Reliability Perspective,* Electric Power Systems Research 34, 1995.

Gates, J., Billinton, R. and Wacker, G., *Electric Service Reliability Worth Evaluation for Governments, Institutions and Office Buildings,* IEEE Transactions on Power Systems, Vol. 14, February 1999.

Goel, L. and Billinton, R., *Monte Carlo Simulation Applied to Distribution Feeder Reliability Evaluation,* Electric Power Systems Research, 1994.

Goel, L., Billinton, R. and Gupta, R., *Effects of Adverse Weather Conditions and Higher Order Outages on Customer Interruption Costs in Electric Sub-transmission Systems,* Electric Power Systems Research 31, 1994.

Goel, L., Liang, X. and Y-Ou, *Monte Carlo Simulation-based Customer Service Reliability Assessment,* Electric Power Systems Research 49, 1999.

Guo Yongji, Xi Yongjian, Xiao Kai and Yang Huiyi, *Composite System Reliability Evaluation-based on Monte Carlo Simulation Combined with Outages Screening,* IEEE Transactions on Power Systems, Vol. 14, No. 2, May 1999.

IEEE Committee Report, *Bulk System Reliability—Predictive Indices,* IEEE Power Engineering Review, November 1990.

Koval, D.U. and Billinton, R., *Statistical and Analytical Evaluation of the Duration and Cost of Customer Interruptions,* IEEE PES Winter Meeting, USA, 1979.

Liang, X. and Goel, L., *Distribution System Reliability Evaluation Using the Monte Carlo Simulation Method,* Electric Power Systems Research 40, 1997.

Nima Amjady and Mehadi Ehsan, *Evaluation of Power Systems Reliability by an Artificial Neural Network,* IEEE Transactions on Power Systems, February, 1999.

Power System Research Group, *Assessment of Reliability Worth in Electric Power Systems in Canada.* Department of Electrical Engineering, Saskatoon, June 1993.

Qiu, J. and Girgis, A.A., *Optimization of Power System Reliability Level by Stochastic Programming,* Electric Power Systems Research, 1993.

Sanghui, A., *Measurement and Application of Customer Interruption Costs/value of Service for Cost-benefit Reliability Evaluation : Some Commonly Raised Issues,* IEEE Transactions on Power Systems, 1990.

Shipley, R.B., Patton, A.D. and Denison, J.S., *Power Reliability Cost versus Worth,* IEEE Transactions on Power Apparatus Systems, PAS-91, 1972.

Simpson, Bob, *Progress Towards Integration of Reliability Concepts in a Segregated Electric Utility System,* Reliability Engineering and System Safety 46, 1994.

Skuf, L.U., *Ontario Hydro Surveys on Power System Reliability : Summary of Customer Viewpoints,* Ontario Hydro Rep. No. R and MR 80-12, EPRI Seminar, October, 1983.

Subramaniam, R.K., Wacker, G. and Billinton, R., *Understanding Commercial Losses Resulting from Electric Service Interruptions,* IEEE Transactions on Industrial Applications, 1993.

Sullivan, M.J., Suddeth, B.N., Vardell, T. and Vojdani, A., *Interruption Costs, Customer Satisfaction and Expectations for Service Reliability,* IEEE Transactions on Power Systems, Vol. II, No. 2, May 1996.

Tollefson, G., Billinton, R. and Wacker, G., *A Canadian Customer Survey to Assess Power System Reliability Worth,* IEEE 93, PWRS, 1993.

Wacker, G. and Billinton, R., *Customer Cost of Electric Interruptions,* IEEE Proc. 77(6), 1989.

Wacker, G., and Tollefson, G., *Electric Power System Customer Interruption Cost Assessment,* Reliability Engineering and Safety 46, 1994.

Wacker, G., Billinton, R. and Brewer, R.E., *Farm Losses Resulting from Electric Service Interruptions,* Canadian Electrical Association, R&D Project 309U403 Report, 1987.

Wacker, G., Tollefson, G. and Billinton, R., *Understanding the Customers' Role in Reliability Worth Assessment Using Customer Surveys,* Inter-RAMQ Conference for the Power Industry, 1992.

Wacker, G., Wojczynski, E. and Billinton, R., *Cost/Benefit Considerations in Providing an Adequate Electric Energy Supply,* Third Symposium on Large Engineering Systems, July 1980.

Wacker, G., Wojczynski, E. and Billinton, R., *Customer Damage Resulting from Electric Service Interruption,* R and D Project 907, U131 Rep., Canadian Electricity Association, 1982.

Wacker, G., Wojczynski, E., and Billinton, R., *Interruption Cost Methodology and Results—A Canadian Residential Survey,* IEEE Transactions on Power Apparatus and Systems, Vol. PAS-102, No. 10, October, 1983.

William G. Zikmund, *Exploring Marketing Research,* The Dryden Press, CBS Publishing Japan Ltd., 1986.

Wojczynski, E. and Billinton, R., *Effects of Distribution System Reliability Index Distribution upon Interruption Cost/Reliability Worth Estimates,* IEEE Transactions on Power Apparatus and Systems, PAS-104, 1985.

Wojczynski, E., Billinton, R. and Wacker, G., *Interruption Cost Methodology and Results—A Canadian Commercial and Small Industry Survey,* IEEE Transactions on Power Apparatus and Systems, Vol. PAS-103, No. 2, Feb. 1984.

Index